# FORREST M. MIMMS III

## ENGINEER'S MINI-NOTEBOOK SERIES

# VOLUME IV: ELECTRONIC FORMULAS, SYMBOLS & CIRCUITS

COPYRIGHT © 1986, 1988, 2000, 2007
BY FORREST M. MIMS III
ALL RIGHTS RESERVED

*Published for Forrest M. Mims III by:*
**Master Publishing, Inc.**
6125 W. Howard Street
Niles, IL 60714
847-763-0916 (voice)
847-763-0918 (fax)
masterpubl@aol.com (e-mail)

*Visit Master Publishing on the Internet at:*
www.masterpublishing.com

*order on line at:*
www.forrestmims.com

**REGARDING THESE BOOK MATERIALS:**
Reproduction, publication or duplication of this book, or any part thereof, in any manner, mechanically, electronically, or photographically is prohibited without the express written permission of the Author and Publisher. For permission and other rights under this Copyright, write Master Publishing.

**The Author, Publisher, and Seller assume no liability with respect to the use of the information contained herein.**

Printed in the United States of America

# I. FORMULAS, TABLES AND BASIC CIRCUITS

# II. SCHEMATIC SYMBOLS, DEVICE PACKAGES, DESIGN AND TESTING

# III. BASIC SEMICONDUCTOR CIRCUITS

# IV. DIGITAL LOGIC CIRCUITS

# I. FORMULAS, TABLES AND BASIC CIRCUITS

## 1. ELECTRONIC FORMULAS

### DIRECT CURRENT

A DIRECT CURRENT (DC) FLOWS IN ONE DIRECTION, EITHER STEADILY OR IN PULSES.

CURRENT (I) - THE QUANITY OF ELECTRONS PASSING A GIVEN POINT. (UNIT: AMPERE)

VOLTAGE (V) - ELECTRICAL PRESSURE OR FORCE (UNIT: VOLT)

RESISTANCE (R) - RESISTANCE TO THE FLOW OF CURRENT. (UNIT. OHM)

POTENTIAL DIFFERENCE - THE DIFFERENCE IN VOLTAGE BETWEEN THE TWO ENDS OF A CONDUCTOR.

### OHM'S LAW

A POTENTIAL DIFFERENCE OF 1 VOLT WILL FORCE A CURRENT OF 1 AMPERE THROUGH A RESISTANCE OF 1 OHM. HERE ARE VARIOUS FORMULAS FOR OHM'S LAW:

$$V = I \times R$$

$$I = \frac{V}{R}$$

$$R = \frac{V}{I}$$

$$P = I \times V \text{ (OR) } I^2 \times R$$

OHM'S LAW HELPER

| V | |
|---|---|
| I | R |

THIS DIAGRAM SHOWS THE RELATIONSHIP OF V, I AND R.

# RESISTOR NETWORKS

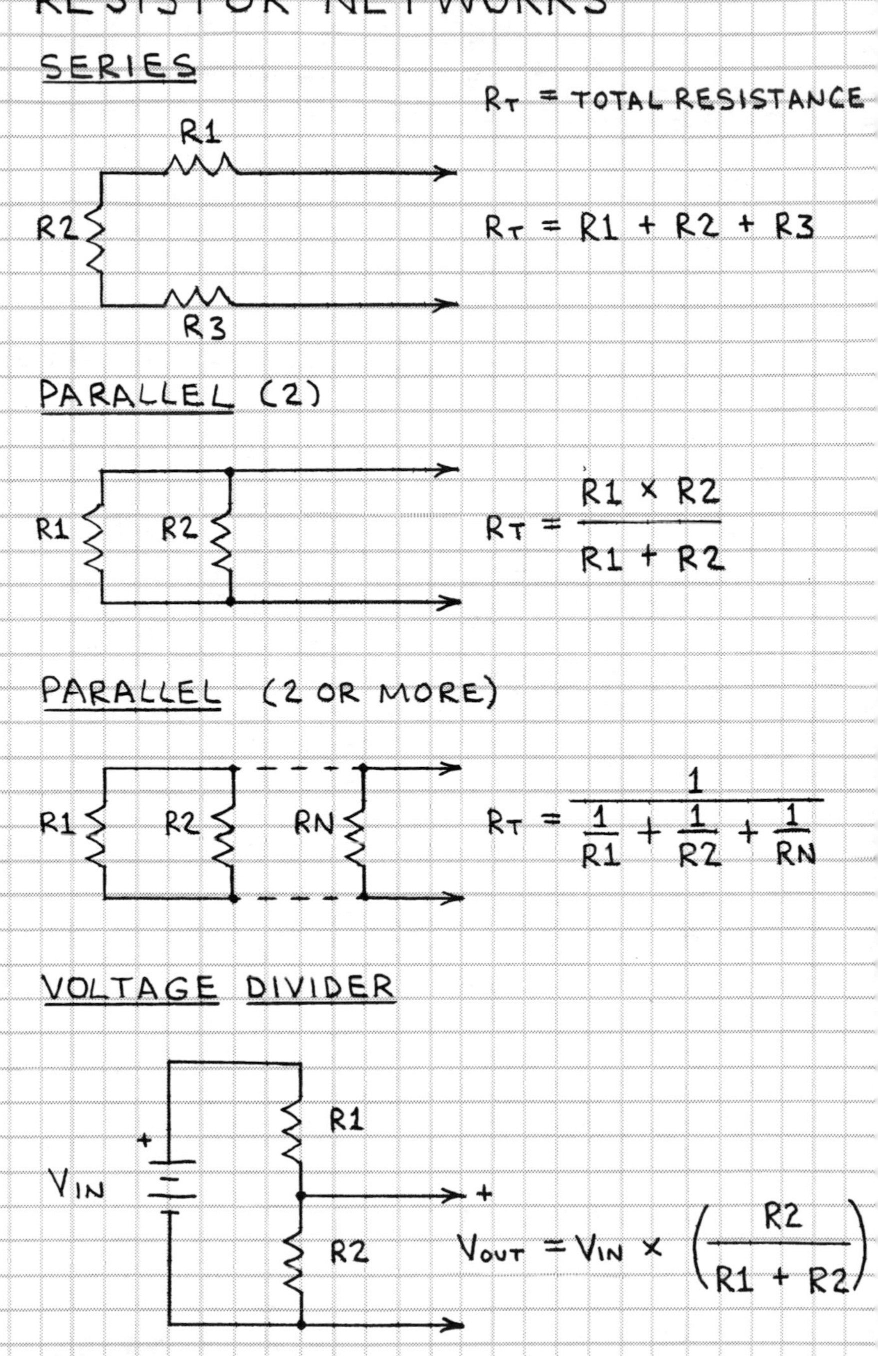


R1 AND R2 CAN BE A POTENTIOMETER.

# ALTERNATING CURRENT

AN ALTERNATING CURRENT (AC) FLOWS IN BOTH DIRECTIONS THROUGH A CONDUCTOR.

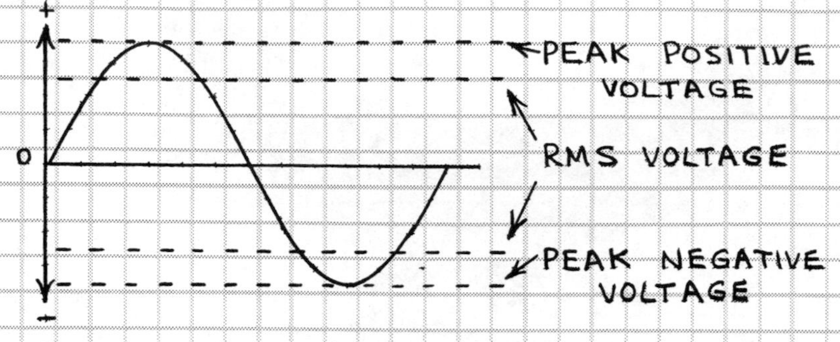


SEE THE DEFINITIONS OF I, V, R AND P ON PAGE 4.

<u>PEAK VOLTAGE</u> — MAXIMUM POSITIVE AND NEGATIVE EXCURSIONS OF AN ALTERNATING CURRENT.

<u>RMS VOLTAGE</u> — (ROOT-MEAN-SQUARE VOLTAGE) THAT AC VOLTAGE THAT EQUALS A DC VOLTAGE THAT DOES THE SAME WORK. FOR A SINE WAVE, 0.707 TIMES THE PEAK VOLTAGE.

<u>IMPEDANCE</u> (Z) — THE OPPOSITION TO AN ALTERNATING CURRENT PRESENTED BY A CIRCUIT. (UNIT: OHM)

AVERAGE AC VOLTAGE = 0.637 × PEAK
= 0.9 × RMS

RMS AC VOLTAGE = 0.707 × PEAK
= 1.11 × AVERAGE

PEAK AC VOLTAGE = 1.414 × RMS
= 1.57 × AVERAGE

# OHM'S LAW

$V = I \times Z$

$I = \frac{E}{Z}$

$Z = \frac{E}{I}$

$P = E \times I \times \cos\theta$

$\theta$ IS PHASE ANGLE, THE DIFFERENCE IN DEGREES BETWEEN CURRENT AND VOLTAGE. CURRENT LEADS VOLTAGE IN A CAPACITIVE CIRCUIT AND LAGS VOLTAGE IN A REACTIVE CIRCUIT. IN A RESISTIVE CIRCUIT $\theta$ IS 0°. THE COSINE OF 0° IS 1. THUS IN A RESISTIVE CIRCUIT $P = E \times I$.

# CAPACITOR NETWORKS

SERIES

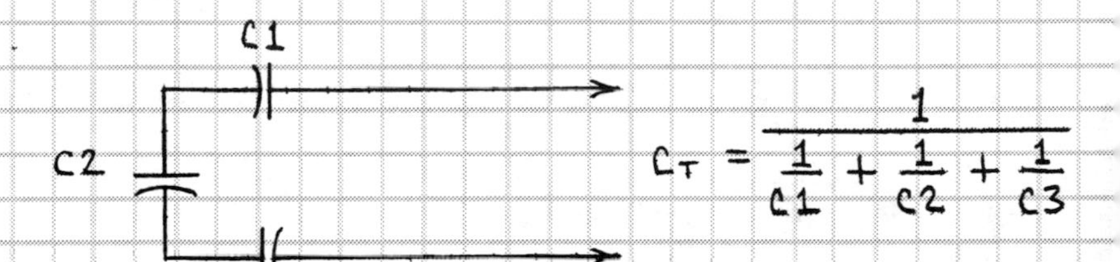


$$C_T = \frac{1}{\frac{1}{C1} + \frac{1}{C2} + \frac{1}{C3}}$$

SERIES

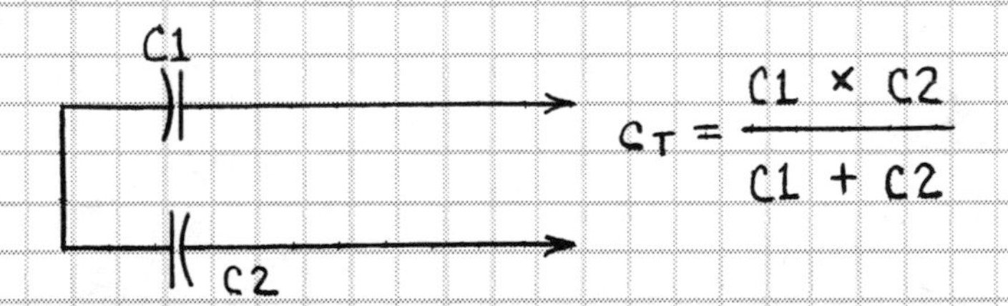


$$C_T = \frac{C1 \times C2}{C1 + C2}$$

PARALLEL (2 OR MORE)

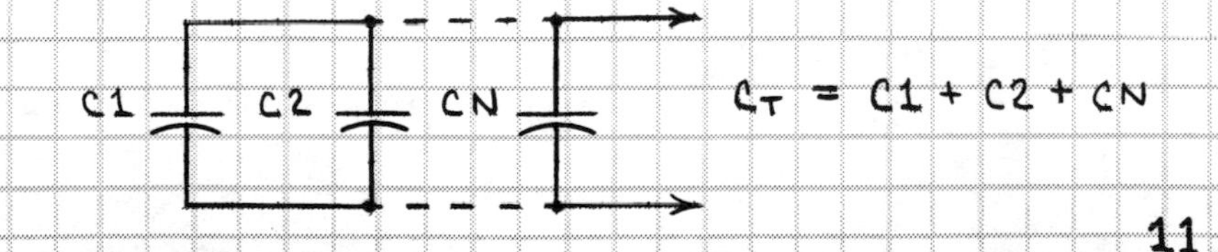


$$C_T = C1 + C2 + CN$$

# 2. MATHEMATICS

## SYMBOLS

| | |
|---|---|
| $+$ | PLUS, POSITIVE OR ADD |
| $-$ | MINUS, NEGATIVE OR SUBTRACT |
| $\times$ OR $*$ | MULTIPLY |
| $\div$ OR $/$ | DIVIDE |
| $=$ | EQUAL(S) |
| $\neq$ | DOES NOT EQUAL |
| $\cong$ | APPROXIMATELY EQUAL |
| $>$ | GREATER THAN |
| $\geq$ | EQUAL TO OR GREATER THAN |
| $<$ | LESS THAN |
| $\leq$ | LESS THAN OR EQUAL TO |
| $\pm$ | PLUS OR MINUS; CHANGE SIGN |
| $1/n$ | RECIPROCAL (1/2 = 0.5) |
| $\sqrt{n}$ | SQUARE ROOT OF n |
| $\sqrt[3]{n}$ | CUBE ROOT OF n |

## POWERS OF TEN

| | | |
|---|---|---|
| $10^{-9}$ | = 0.000000001 | 1 BILLIONTH (NANO) |
| $10^{-8}$ | = 0.00000001 | |
| $10^{-7}$ | = 0.0000001 | |
| $10^{-6}$ | = 0.000001 | 1 MILLIONTH (MICRO) |
| $10^{-5}$ | = 0.00001 | |
| $10^{-4}$ | = 0.0001 | |
| $10^{-3}$ | = 0.001 | 1 THOUSANDTH (MILLI) |
| $10^{-2}$ | = 0.01 | |
| $10^{-1}$ | = 0.1 | |
| $10^{0}$ | = 1 | 1 UNIT |
| $10^{1}$ | = 10 | |
| $10^{2}$ | = 100 | |
| $10^{3}$ | = 1,000 | THOUSAND (KILO) |
| $10^{4}$ | = 10,000 | |
| $10^{5}$ | = 100,000 | |
| $10^{6}$ | = 1,000,000 | MILLION (MEGA) |
| $10^{7}$ | = 10,000,000 | |
| $10^{8}$ | = 100,000,000 | |
| $10^{9}$ | = 1,000,000,000 | BILLION (GIGA) |

# ALGEBRAIC TRANSPOSITION

IF $A + B = C$, THEN:

$$A = C - B$$

$$B = C - A$$

$$A + B - C = 0$$

IF $A = \frac{B}{C}$, THEN

$$B = AC$$

$$C = \frac{B}{A}$$

IF $\frac{A}{B} = \frac{C}{D}$, THEN:

$$AD = BC$$

$$A = \frac{BC}{D}$$

$$B = \frac{AD}{C}$$

$$C = \frac{AD}{B}$$

$$D = \frac{BC}{A}$$

# LAW OF EXPONENTS

$$\left(\frac{a}{b}\right)^x = \frac{a^x}{b^x} \qquad (a^x)(a^y) = a^{x+y}$$

$$\frac{a^x}{a^y} = a^{x-y} \qquad (a^x)^y = a^{xy}$$

$$a^{-x} = \frac{1}{a^x} \qquad a^{\frac{x}{y}} = \sqrt[y]{a^x}$$

# COMMON LOGARITHMS

THE COMMON LOGARITHM ($LOG_{10}$ OR LOG) OF A NUMBER IS THE POWER OF 10 THAT EQUALS THE NUMBER. SINCE $10^2 = 100$, 2 IS THE LOG OF 100. THE ANTILOGARITHM (ANTILOG) IS THE NUMBER THAT EQUALS A LOGARITHM. THUS THE ANTILOG OF 2 IS 100. THE LOG OF NUMBERS GREATER THAN 1 IS POSITIVE; THE LOG OF NUMBERS LESS THAN 1 IS NEGATIVE. THUS THE LOG OF $10^{-2}$ OR 0.01 IS −2. $A \times B$ = ANTILOG (LOG A + LOG B); $A \div B$ = ANTILOG (LOG A − LOG B). SCIENTIFIC CALCULATORS HAVE LOG AND ANTILOG KEYS.

# THE DECIBEL

THE DECIBEL (dB) IS A UNIT OF MEASURE THAT PERMITS TWO DIFFERENT SIGNALS TO BE COMPARED ON A LOGARITHMIC SCALE. THE SENSITIVITY OF RECEIVERS AND THE GAIN OF AMPLIFIERS ARE OFTEN GIVEN IN DECIBELS. THE DIFFERENCE IN dB BETWEEN THE POWER OF A SIGNAL AT THE INPUT OF AN AMPLIFIER (P1) AND THE POWER OF THE AMPLIFIER'S OUTPUT (P2) IS:

$$dB = 10 \text{ LOG } (P2/P1)$$

THE DIFFERENCE IN dB BETWEEN THE VOLTAGE (V) AND CURRENT (I) AT THE INPUT (V1 AND I1) AND OUTPUT (V2 AND I2) OF AN AMPLIFIER IS:

$$dB = 20 \text{ LOG } (V2/V1)$$

$$dB = 20 \text{ LOG } (I2/I1)$$

NOTE THAT DECIBELS DEFINE THE <u>RATIO</u> BETWEEN TWO SIGNAL LEVELS, NOT THEIR ABSOLUTE VALUE.

EXAMPLE: DETERMINE THE VOLTAGE GAIN IN dB OF THIS OPERATIONAL AMPLIFIER.

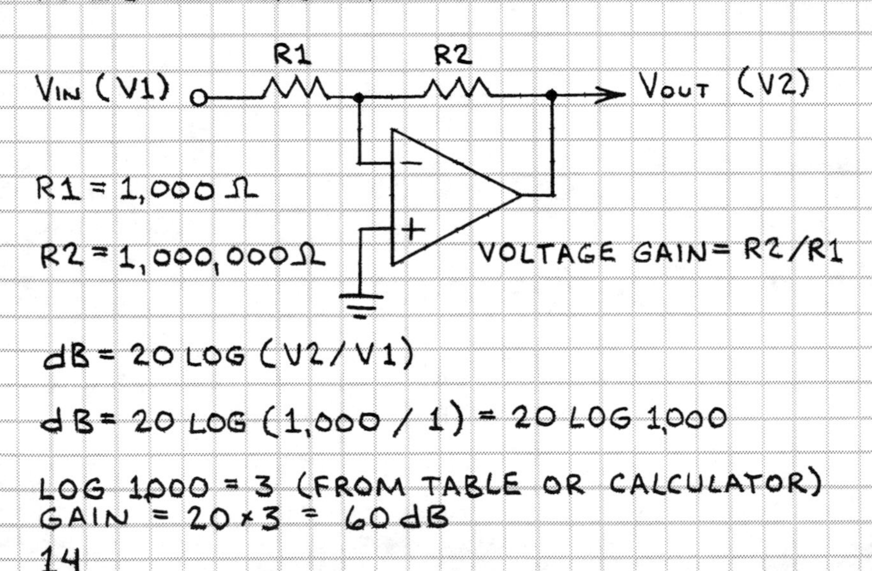


$$dB = 20 \text{ LOG } (V2/V1)$$

$$dB = 20 \text{ LOG } (1{,}000 / 1) = 20 \text{ LOG } 1{,}000$$

LOG 1000 = 3 (FROM TABLE OR CALCULATOR)
GAIN = 20 × 3 = 60 dB

# DECIBEL (dB) TABLE

| − | | | + | |
|---|---|---|---|---|
| VOLTAGE OR CURRENT RATIO | POWER RATIO | dB | VOLTAGE OR CURRENT RATIO | POWER RATIO |
| 1.0000 | 1.0000 | 0 | 1.0000 | 1.0000 |
| .8913 | .7943 | 1 | 1.1220 | 1.2589 |
| .7943 | .6310 | 2 | 1.2589 | 1.5849 |
| .7079 | .5012 | 3 | 1.4125 | 1.9953 |
| .6310 | .3981 | 4 | 1.5849 | 2.5119 |
| .5623 | .3162 | 5 | 1.7783 | 3.1623 |
| .5012 | .2512 | 6 | 1.9953 | 3.9811 |
| .4467 | .1995 | 7 | 2.2387 | 5.0119 |
| .3981 | .1585 | 8 | 2.5119 | 6.3096 |
| .3548 | .1259 | 9 | 2.8184 | 7.9433 |
| .3162 | .1000 | 10 | 3.1623 | 10.000 |
| .1000 | .0100 | 20 | 10.000 | 100.00 |
| .0316 | .0010 | 30 | 31.623 | 1,000.0 |
| .0100 | .0001 | 40 | 100.00 | 10,000 |
| .0032 | .00001 | 50 | 316.23 | 100,000 |
| .0010 | $10^{-6}$ | 60 | 1,000.0 | $10^{6}$ |
| .0003 | $10^{-7}$ | 70 | 3,162.3 | $10^{7}$ |
| .0001 | $10^{-8}$ | 80 | 10,000 | $10^{8}$ |
| .00003 | $10^{-9}$ | 90 | 31,623 | $10^{9}$ |
| .00001 | $10^{-10}$ | 100 | 100,000 | $10^{10}$ |

# POWER-dBm EQUIVALENTS

RECEIVER SENSITIVITY IS OFTEN GIVEN IN dB WITH RESPECT TO 1 MILLIWATT.

| dBm | POWER (mW) | UNITS |
|---|---|---|
| 10 | 10.000000 | 10 MILLIWATTS |
| 0 | 1.000000 | 1 MILLIWATT |
| -10 | .100000 | 100 MICROWATTS |
| -20 | .010000 | 10 MICROWATTS |
| -30 | .001000 | 1 MICROWATT |
| -40 | .000100 | 100 NANOWATTS |
| -50 | .000010 | 10 NANOWATTS |
| -60 | .000001 | 1 NANOWATT |

# NUMBER SYSTEMS

A NUMBER SYSTEM CAN BE BASED ON ANY NUMBER OF DIGITS. THE COMMON DECIMAL SYSTEM HAS 10 DIGITS. THE BINARY SYSTEM HAS 2 DIGITS; THE HEXADECIMAL SYSTEM HAS 16 DIGITS. NUMBERS ARE WRITTEN AS SUCCESSIVE POWERS OF THE BASE OF THE NUMBER SYSTEM. THUS:

$4327_{10}$

$$7 \times 10^0 = 7 \times 1 = 7$$
$$2 \times 10^1 \quad 2 \times 10 = 20$$
$$3 \times 10^2 \quad 3 \times 100 = 300$$
$$4 \times 10^3 = 4 \times 1000 = \underline{4000}$$
$$4327$$

# BINARY NUMBERS

IN ELECTRONIC CIRCUITS DECIMAL NUMBERS ARE USUALLY REPRESENTED BY BINARY NUMBERS. BINARY NUMBERS ALSO SERVE AS CODES THAT REPRESENT LETTERS OF THE ALPHABET, VOLTAGES, COMPUTER INSTRUCTIONS, ETC. A BINARY 0 OR 1 IS A <u>BIT</u>. A PATTERN OF 4 BITS IS A <u>NIBBLE</u>. A PATTERN OF 4 BITS IS A <u>BYTE</u> OR <u>WORD</u>.

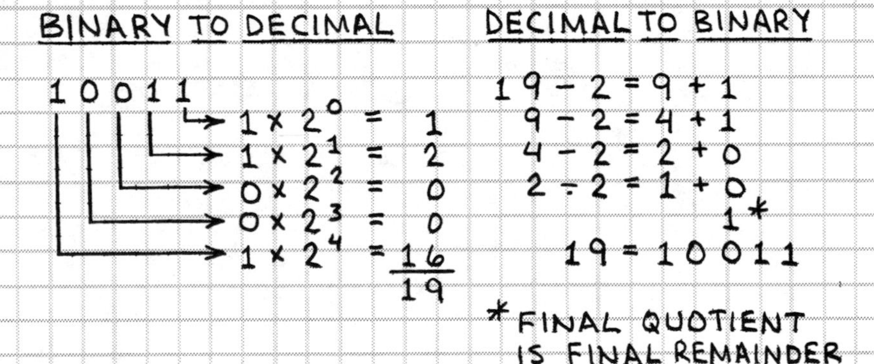


<u>BINARY CODED DECIMAL</u> (BCD). A SYSTEM IN WHICH EACH DECIMAL DIGIT IS ASSIGNED ITS BINARY EQUIVALENT (19 = 0001 1001).

# NUMBER SYSTEM EQUIVALENTS

DEC (DECIMAL) BIN (BINARY)
BCD (BINARY CODED DECIMAL) HEX (HEXADECIMAL)

| DEC | BIN | BCD | HEX |
|---|---|---|---|
| 0 | 0 | 0000 0000 | 0 |
| 1 | 1 | 0000 0001 | 1 |
| 2 | 10 | 0000 0010 | 2 |
| 3 | 11 | 0000 0011 | 3 |
| 4 | 100 | 0000 0100 | 4 |
| 5 | 101 | 0000 0101 | 5 |
| 6 | 110 | 0000 0110 | 6 |
| 7 | 111 | 0000 0111 | 7 |
| 8 | 1000 | 0000 1000 | 8 |
| 9 | 1001 | 0000 1001 | 9 |
| 10 | 1010 | 0001 0000 | A |
| 11 | 1011 | 0001 0001 | B |
| 12 | 1100 | 0001 0010 | C |
| 13 | 1101 | 0001 0011 | D |
| 14 | 1110 | 0001 0100 | E |
| 15 | 1111 | 0001 0101 | F |
| 16 | 10000 | 0001 0110 | 10 |
| 17 | 10001 | 0001 0111 | 11 |
| 18 | 10010 | 0001 1000 | 12 |
| 19 | 10011 | 0001 1001 | 13 |
| 20 | 10100 | 0010 0000 | 14 |
| 21 | 10101 | 0010 0001 | 15 |
| 22 | 10110 | 0010 0010 | 16 |
| 23 | 10111 | 0010 0011 | 17 |
| 24 | 11000 | 0010 0100 | 18 |
| 25 | 11001 | 0010 0101 | 19 |
| 26 | 11010 | 0010 0110 | 1A |
| 27 | 11011 | 0010 0111 | 1B |
| 28 | 11100 | 0010 1000 | 1C |
| 29 | 11101 | 0010 1001 | 1D |
| 30 | 11110 | 0011 0000 | 1E |
| 31 | 11111 | 0011 0001 | 1F |
| 32 | 100000 | 0011 0010 | 20 |
| 64 | 1000000 | 0110 0100 | 40 |
| 96 | 1100000 | 1001 0110 | 60 |
| 99 | 1100011 | 1001 1001 | 63 |

# 3. CONSTANTS AND STANDARDS

## U.S. WEIGHTS AND MEASURES

### LINEAR

1,000 MILS = 1 INCH (IN)
12 INCHES = 1 FOOT (FT)
3 FT = 1 YARD (YD)
5,280 FT = 1 MILE (MI)

### AREA

1 FOOT$^2$ = 144 IN$^2$
1 YARD$^2$ = 9 FT$^2$
1 ACRE = 43,560 FT$^2$
1 MILE$^2$ = 640 ACRES

### VOLUME

1 FOOT$^3$ = 1,728 IN$^3$
1 YARD$^3$ = 27 FEET$^3$

### MASS

16 OUNCES (OZ) = 1 POUND (lb)

## METRIC WEIGHTS AND MEASURES

### LINEAR

1,000 MICROMETERS ($\mu$m) = 1 MILLIMETER (mm)
10 mm = 1 CENTIMETER (cm)
100 cm = 1 METER (m)
1,000 METERS = 1 KILOMETER (KM)

### AREA

100 mm$^2$ = 1 cm$^2$
10,000 cm$^2$ = 1 m$^2$

### VOLUME

1 cm$^3$ = 1 MILLILITER (ml)
1,000 ml = 1 LITER (l)

### MASS

1,000 MILLIGRAMS (mg) = 1 gram (g)

# U.S.-METRIC CONVERSION

| TO CONVERT | INTO | MULTIPLY BY |
|---|---|---|
| MICROMETERS | MILS | $3.937 \times 10^{-2}$ |
| MILS | MICROMETERS | 25.4 |
| MILLIMETERS | MILS | 39.37 |
| MILS | MILLIMETERS | $2.54 \times 10^{-2}$ |
| MILLIMETERS | INCHES | $3.937 \times 10^{-2}$ |
| INCHES | MILLIMETERS | 25.4 |
| CENTIMETERS | INCHES | 0.3937 |
| INCHES | CENTIMETERS | 2.54 |
| INCHES | METERS | $2.54 \times 10^{-2}$ |
| METERS | INCHES | 39.37 |
| FEET | METERS | $30.48 \times 10^{-2}$ |
| METERS | FEET | 3.281 |
| METERS | YARDS | 1.094 |
| YARDS | METERS | 0.9144 |
| KILOMETERS | FEET | 3281 |
| FEET | KILOMETERS | $3.408 \times 10^{-4}$ |
| KILOMETERS | MILES | 0.6214 |
| MILES | KILOMETERS | 1.609 |
| GRAMS | OUNCES | $3.527 \times 10^{-2}$ |
| OUNCES | GRAMS | 28.3495 |
| KILOGRAMS | POUNDS | 2.205 |
| POUNDS | KILOGRAMS | 0.4536 |

# FAMILIAR EXAMPLES

## DIMENSIONS

DIME ≈ 1 mm × 1.8 cm
NICKEL ≈ 2 mm × 2.1 cm
QUARTER ≈ 2 mm × 2.4 cm
1-MIL PLASTIC FILM = 25.4 μm

## MASS

PLASTIC TO-92 TRANSISTOR ≈ 0.25 g
8-PIN MINI DIP IC ≈ 0.5 g
16-PIN DIP IC ≈ 1 05 g
NICKEL ≈ 5 g

# TEMPERATURE

$°\text{FAHRENHEIT} = (°\text{CELSIUS} \times \frac{9}{5}) + 32 = °F$

$°\text{CELSIUS} = \frac{5}{9} \times (°\text{FAHRENHEIT} - 32) = °C$

| | °C | °F |
|---|---|---|
| LEAD MELTS → | 328 | 622.4 |
| WATER BOILS → | 100 | 212 |
| | 90 | 194 |
| | 80 | 176 |
| | 70 | 158 |
| | 60 | 140 |
| | 50 | 122 |
| | 40 | 104 |
| HUMAN BODY (37°C; 98 6°F) | | |
| | 30 | 86 |
| ROOM TEMPERATURE (22°C) | 20 | 68 |
| | 10 | 50 |
| WATER FREEZES → | 0 | 32 |

TYPICAL SEMICONDUCTOR OPERATING TEMPERATURE RANGE:

COMMERCIAL: 0° TO 70°C
INDUSTRIAL : -65° TO 150°C

# SOLDER

THE MOST COMMON ELECTRONIC SOLDER IS 60/40 (60% TIN AND 40% LEAD). ITS MELTING POINT IS 183° TO 190°C (361° TO 374°F).

# COPPER WIRE

| AWG | DIA | OHMS PER 1000 FT | FT PER POUND |
|---|---|---|---|
| 10 | 101.9 | .9989 | 31.82 |
| 12 | 80.8 | 1.588 | 50.59 |
| 14 | 64.1 | 2.525 | 80.44 |
| 16 | 50.8 | 4.016 | 127.9 |
| 18 | 40.3 | 6.385 | 203.4 |
| 20 | 32 0 | 10.15 | 323.4 |
| 22 | 25.4 | 16.14 | 514.2 |
| 24 | 20.1 | 25.67 | 817.7 |
| 26 | 159 | 40.81 | 1,300.0 |
| 28 | 12.6 | 64.90 | 2,067.0 |
| 30 | 10.0 | 103.2 | 3,287.0 |
| 32 | 7.9 | 164.1 | 5,227.0 |
| 34 | 6.3 | 260.9 | 8,310.0 |
| 36 | 5.0 | 414.8 | 13,210.0 |
| 38 | 4.0 | 659.6 | 21,010.0 |
| 40 | 3.1 | 1,049.0 | 33,410.0 |

AWG – AMERICAN WIRE GAUGE
DIA – DIAMETER IN MILS
OHMS PER 1000 FT – 20° C (68° F)

# RELATIVE RESISTANCES

| | |
|---|---|
| SILVER | 0.936 |
| COPPER | 1.000 |
| GOLD | 1.403 |
| CHROMIUM | 1.530 |
| ALUMINUM | 1.549 |
| TUNGSTEN | 3.203 |
| BRASS | 4.822 |
| PHOSPHOR-BRONZE | 5.533 |
| NICKEL | 5.786 |
| IRON | 5799 |
| TIN | 6.702 |
| STEEL | 9.932 |
| LEAD | 12.922 |
| STAINLESS STEEL | 52.941 |
| NICHROME | 65.092 |

RESISTANCE RELATIVE TO COPPER. 1 FOOT OF CIRCULAR COPPER WIRE 1 MIL IN DIAMETER HAS A RESISTANCE OF 10.37 OHMS. ALTERNATIVELY, COPPER WIRE HAS A RESISTANCE OF 10.37 OHMS PER CIRCULAR MIL FOOT.

# AUDIO FREQUENCY SPECTRUM

MECHANICAL VIBRATION IN SOLIDS, FLUIDS AND GASES PRODUCES WHAT THE BRAIN PERCEIVES AS SOUND.

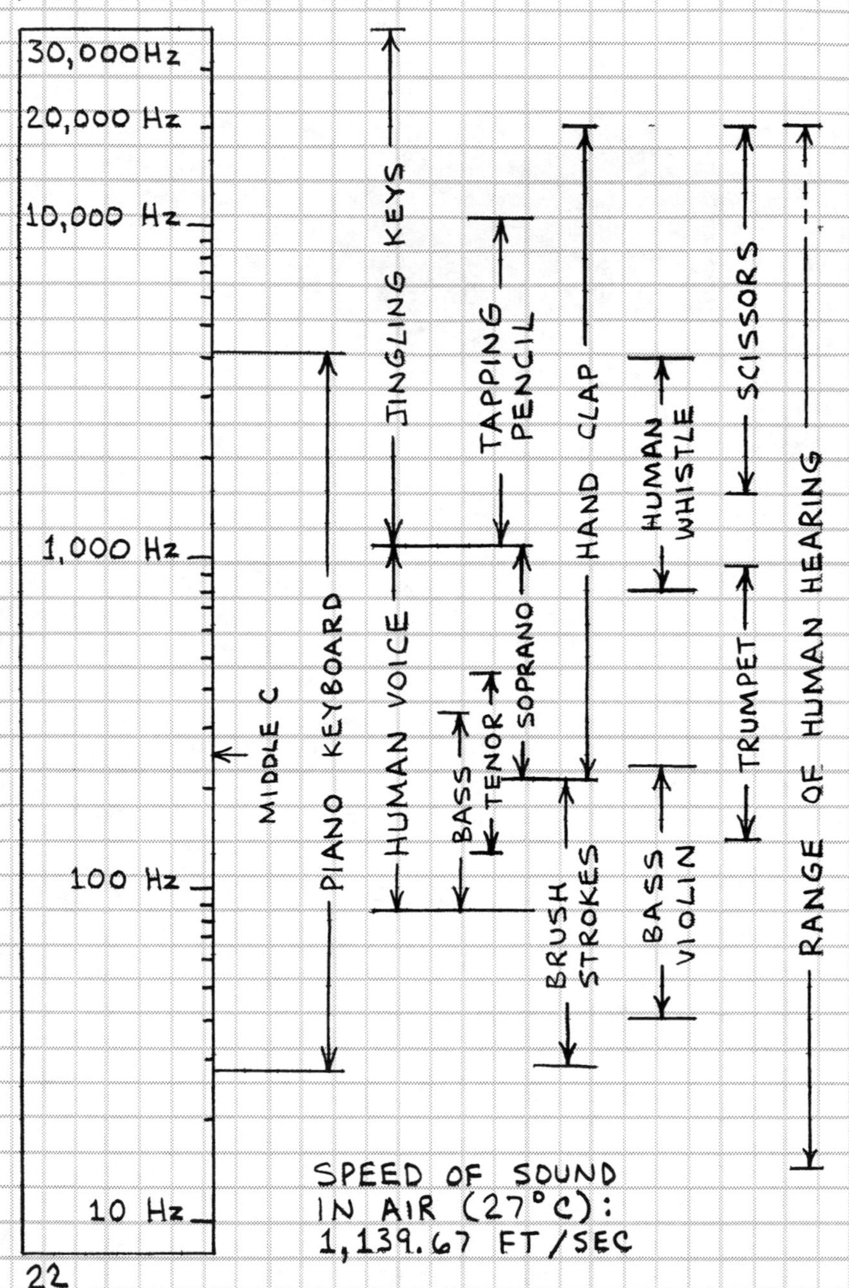

# SOUND INTENSITY LEVELS

| SOUND SOURCE (DISTANCE FROM OBSERVER) | LEVEL (dB) |
|---|---|
| THRESHOLD OF PAIN | 120+ |
| AIRCRAFT ENGINE (20') | 120+ |
| AMPLIFIED ROCK MUSIC | 110 |
| THUNDER | 110 |
| PIEZOELECTRIC BUZZER (12") | 108 |
| AIR FORCE T-38 (2,500' OVERHEAD) | 90 |
| $CO_2$ PELLET GUN (12") | 90 |
| DIGITAL ALARM CLOCK (12") | 85 |
| ELECTRIC TYPEWRITER (18") | 80 |
| AIR FORCE T-38 (1 MILE) | 70 |
| TYPICAL CONVERSATION | 65 |
| PAPER CLIP DROPPED ON DESK (12") | 62 |
| TELEPHONE DIAL TONE (1") | 56 |
| PENCIL ERASER TAPPED ON DESK (12") | 54 |
| COMPUTER KEYBOARD (18") | 61 |
| AVERAGE RESIDENCE | 45 |
| SOFT BACKGROUND MUSIC | 30 |
| QUIET WHISPER | 20 |
| THRESHOLD OF HEARING | 0 |

# ELECTROMAGNETIC SPECTRUM

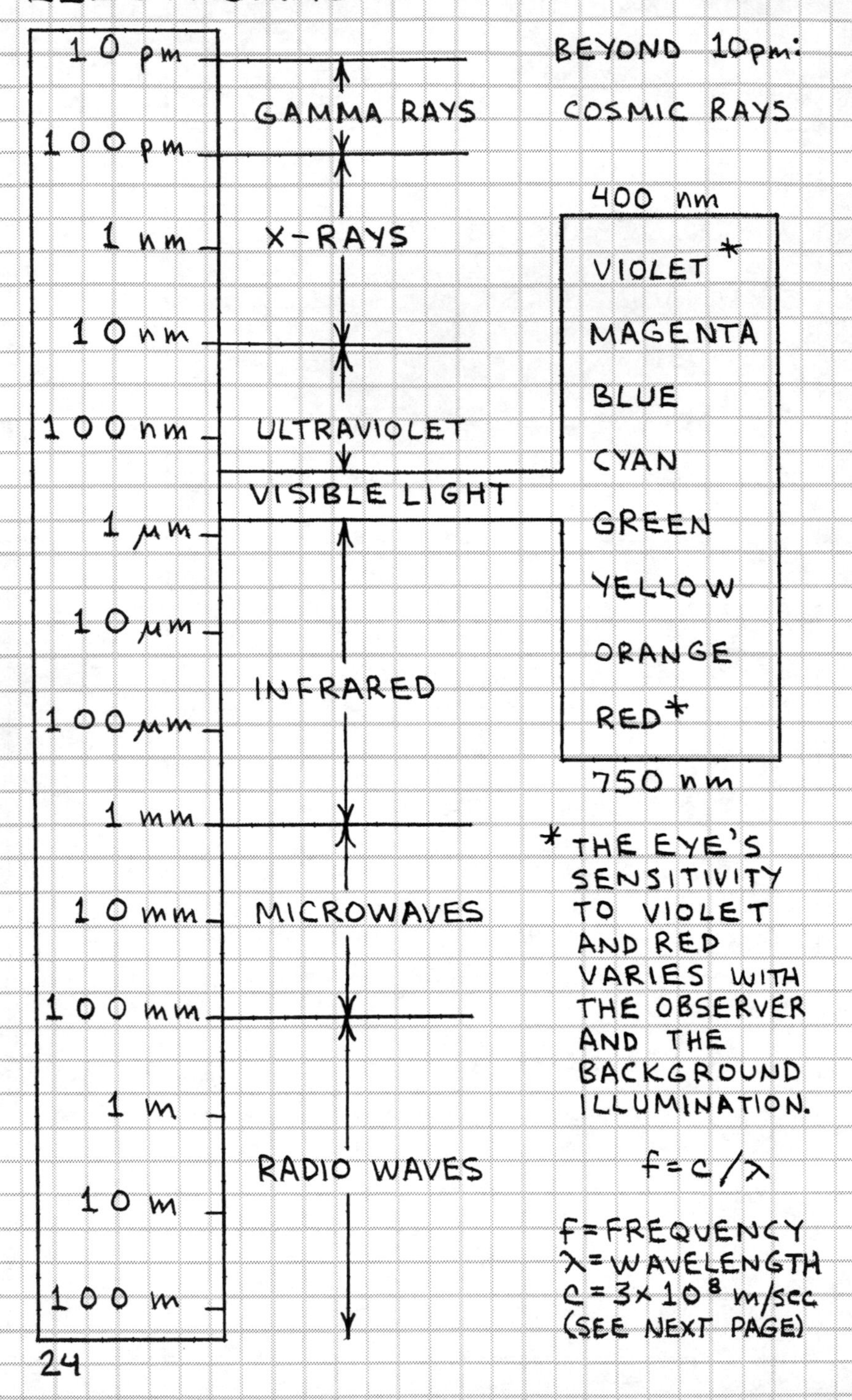


* THE EYE'S SENSITIVITY TO VIOLET AND RED VARIES WITH THE OBSERVER AND THE BACKGROUND ILLUMINATION.

$f = c/\lambda$

f = FREQUENCY
$\lambda$ = WAVELENGTH
$c = 3 \times 10^8$ m/sec

(SEE NEXT PAGE)

# RADIO FREQUENCY SPECTRUM

| FREQUENCY | CLASSIFICATION |
|---|---|
| 3-30 KHz | VERY LOW FREQUENCIES (VLF) |
| 30-300 KHz | LOW FREQUENCIES (LF) |
| 300-3000 KHz | MEDIUM FREQUENCIES (MF) |
| 3-30 MHz | HIGH FREQUENCIES (HF) |
| 30-300 MHz | VERY HIGH FREQUENCIES (VHF) |
| 300-3000 MHz | ULTRA HIGH FREQUENCIES (UHF) |
| 3-30 GHz | SUPER HIGH FREQUENCIES (SHF) |
| 30-300 GHz | EXTREMELY HIGH FREQUENCIES (EHF) |
| 300-3000 GHz | MICROWAVE FREQUENCIES |

# FREQUENCY VS. WAVELENGTH

$$\lambda = \frac{c}{f} \qquad f = \frac{c}{\lambda}$$

$\lambda$ - WAVELENGTH (METERS)
C - SPEED OF LIGHT ($3 \times 10^8$ METERS/SEC)
F - FREQUENCY (HERTZ)

EXAMPLE: THE WAVELENGTH OF A 108 MHz SIGNAL IS $3 \times 10^8 / 1.08 \times 10^6$ OR 2.78 METERS.

# IMPORTANT FREQUENCIES (MHz)

| | |
|---|---|
| .15 - .54: | NAVIGATION BEACONS |
| .5 : | INTERNATIONAL DISTRESS |
| .54 - 1.6: | AM BROADCAST BAND |
| 1.61: | AIRPORT INFORMATION |
| 1.8 - 2.0: | 160 METER AMATEUR BAND |
| 2.3 - 2.498: | 120 METER INT. BROADCAST |
| 2.5: | WWV TIME SIGNAL |
| 3.5 - 4.0. | 80 METER AMATEUR BAND |
| 5.0: | WWV TIME SIGNAL |
| 5.95 - 6.2: | 49 METER INT. BROADCAST |
| 6.2 - 6.525: | MARITIME COMMUNICATIONS |
| 7.0 - 7.3: | 40 METER AMATEUR |
| 7.0 - 7.3: | 40 METER INT. BROADCAST |
| 9.5 - 9.9: | 31 METER INT. BROADCAST |
| 10.0: | WWV TIME SIGNAL |
| 10.1 - 10.15: | 30 METER AMATEUR BAND |
| 10.15 - 11.175. | INT BROADCAST |
| 11.7 - 11.975: | 25 METER INT. BROADCAST |
| 14.0 - 14.35: | 20 METER AMATEUR BAND |
| 15.0: | WWV TIME SIGNAL |
| 20.0 | WWV TIME SIGNAL |
| 21.0 - 21.45 | 15 METER AMATEUR BAND |
| 21.45 - 21.85 | 13 METER INT. BROADCAST |
| 24.89 - 24.99: | 12 METER AMATEUR BAND |
| 25.67 - 26.1: | 11 METER INT. BROADCAST |
| 26.9 - 27.4: | CITIZENS BAND |
| 28.0 - 29.7: | 10 METER AMATEUR BAND |
| 49.82 - 49.9: | LOW POWER COMMUNICATIONS |
| 50.0 - 54.0: | 6 METER AMATEUR BAND |
| 54.0 - 88.0: | TELEVISION (CH. 2-6) |
| 72.03 - 72.9: | RADIO CONTROL (AIRCRAFT ONLY) |
| 75.43 - 75.87: | RADIO CONTROL |
| 88.0 - 108.0: | FM BROADCAST BAND |
| 88.0 - 108.0: | WIRELESS MICROPHONES |
| 108.0 - 118.0: | AIR NAVIGATION BEACONS |
| 118.0 - 136.0: | AIRCRAFT |
| 153 - 155: | POLICE, FIRE, MUNICIPAL |
| 158 - 159 | POLICE, FIRE, MUNICIPAL |
| 162.4 - 162.55: | NOAA WEATHER |
| 174 - 216: | TELEVISION (CH. 7-13) |
| 470 - 890: | TELEVISION (CH. 14-83) |

# TIME CONVERSIONS

| UTC | PST | MST | CST | EST | AST |
|---|---|---|---|---|---|
| 0000 | 4 PM | 5 PM | 6 PM | 7 PM | 8 PM |
| 0100 | 5 PM | 6 PM | 7 PM | 8 PM | 9 PM |
| 0200 | 6 PM | 7 PM | 8 PM | 9 PM | 10 PM |
| 0300 | 7 PM | 8 PM | 9 PM | 10 PM | 11 PM |
| 0400 | 8 PM | 9 PM | 10 PM | 11 PM | MIDNT |
| 0500 | 9 PM | 10 PM | 11 PM | MIDNT | 1 AM |
| 0600 | 10 PM | 11 PM | MIDNT | 1 AM | 2 AM |
| 0700 | 11 PM | MIDNT | 1 AM | 2 AM | 3 AM |
| 0800 | MIDNT | 1 AM | 2 AM | 3 AM | 4 AM |
| 0900 | 1 AM | 2 AM | 3 AM | 4 AM | 5 AM |
| 1000 | 2 AM | 3 AM | 4 AM | 5 AM | 6 AM |
| 1100 | 3 AM | 4 AM | 5 AM | 6 AM | 7 AM |
| 1200 | 4 AM | 5 AM | 6 AM | 7 AM | 8 AM |
| 1300 | 5 AM | 6 AM | 7 AM | 8 AM | 9 AM |
| 1400 | 6 AM | 7 AM | 8 AM | 9 AM | 10 AM |
| 1500 | 7 AM | 8 AM | 9 AM | 10 AM | 11 AM |
| 1600 | 8 AM | 9 AM | 10 AM | 11 AM | 12 AM |
| 1700 | 9 AM | 10 AM | 11 AM | 12 AM | 1 PM |
| 1800 | 10 AM | 11 AM | 12 AM | 1 PM | 2 PM |
| 1900 | 11 AM | 12 AM | 1 PM | 2 PM | 3 PM |
| 2000 | 12 AM | 1 PM | 2 PM | 3 PM | 4 PM |
| 2100 | 1 PM | 2 PM | 3 PM | 4 PM | 5 PM |
| 2200 | 2 PM | 3 PM | 4 PM | 5 PM | 6 PM |
| 2300 | 3 PM | 4 PM | 5 PM | 6 PM | 7 PM |

UTC – COORDINATED UNIVERSAL TIME
(GREENWICH MERIDIAN TIME, LONDON)

PST – PACIFIC STANDARD TIME

MST – MOUNTAIN STANDARD TIME

CST – CENTRAL STANDARD TIME

EST – EASTERN STANDARD TIME

AST – ATLANTIC STANDARD TIME

DAYLIGHT SAVINGS TIME – ADD 1 HOUR

# THE SINE WAVE

THE SINE OR SINUSOIDAL WAVE IS THE MOST COMMON PERIODIC WAVE IN ANALOG ELECTRONIC CIRCUITS. IF PEAK AMPLITUDES ARE +1 AND −1, THEN:

| ANGLE (a) | AMPLITUDE (SIN a) |
|---|---|
| 0° | 0 |
| 30° | 0.500 |
| 45° | 0.707 |
| 90° | 1 |
| 135° | 0.707 |
| 180° | 0 |
| 225° | −0.707 |
| 270° | −1 |
| 315° | −0.707 |
| 360° | 0 |

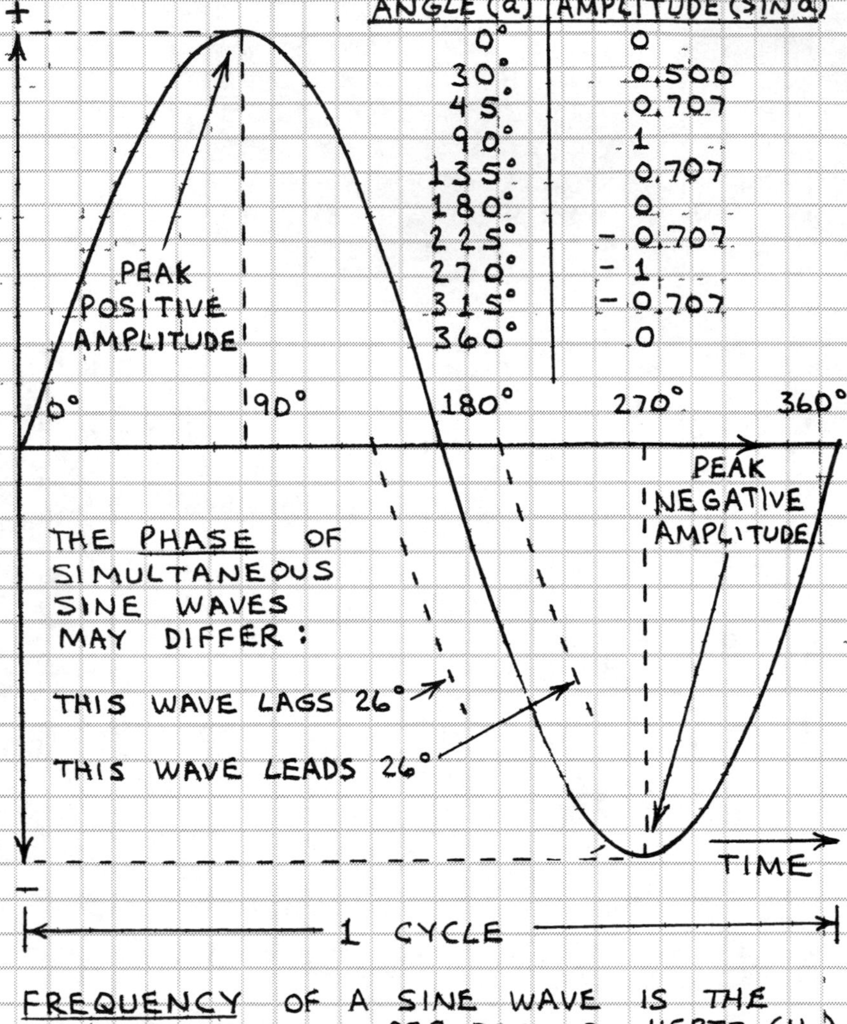

FREQUENCY OF A SINE WAVE IS THE NUMBER OF CYCLES PER SECOND. HERTZ (Hz) IS THE UNIT OF FREQUENCY. ONE HERTZ (1 Hz) IS ONE CYCLE PER SECOND (1 CPS).

PERIOD OF A SINE WAVE IS THE TIME FOR ONE COMPLETE CYCLE TO OCCUR.

# PERIODIC WAVES

MANY DIFFERENT PERIODIC WAVEFORMS CAN BE PROCESSED OR GENERATED BY ANALOG ELECTRONIC CIRCUITS. THEY INCLUDE:

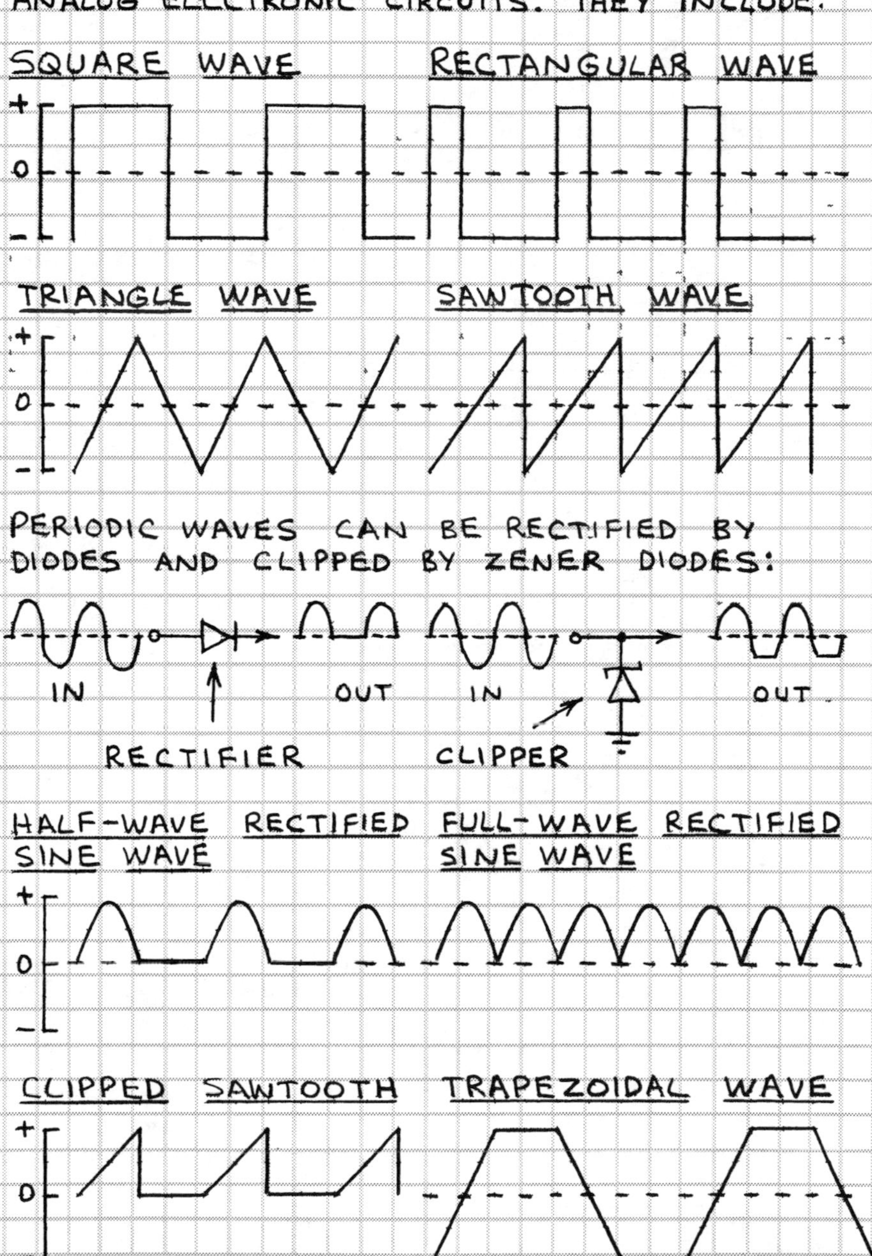

# PULSES

SINGLE PULSES OR TRAINS OF PERIODIC PULSES ARE PROCESSED AND GENERATED BY DIGITAL ELECTRONIC CIRCUITS. THEY ARE ALSO USED TO TRIGGER (ACTIVATE) MANY KINDS OF CIRCUITS.

## THE IDEAL PULSE

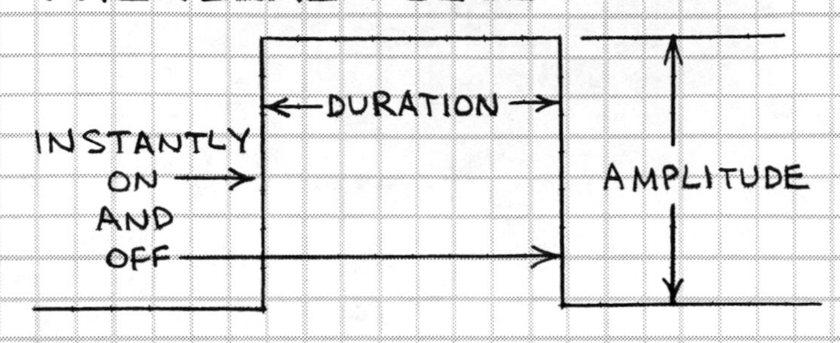


## A REAL PULSE

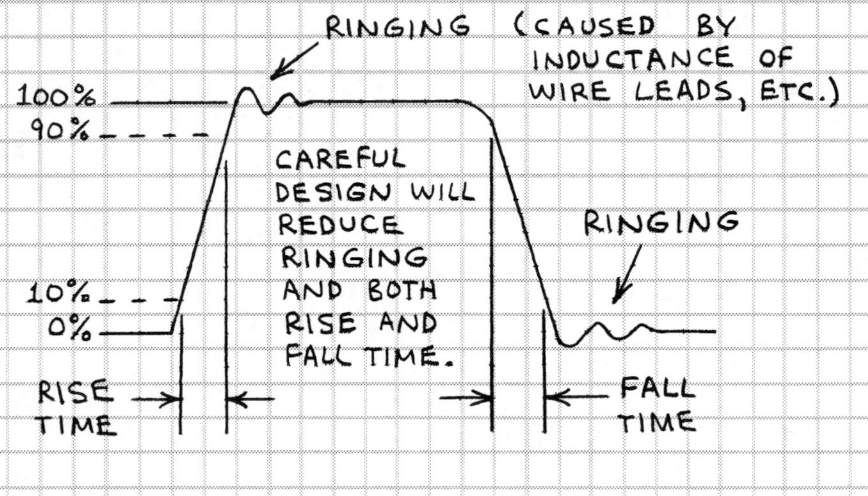


## PULSE TRAIN

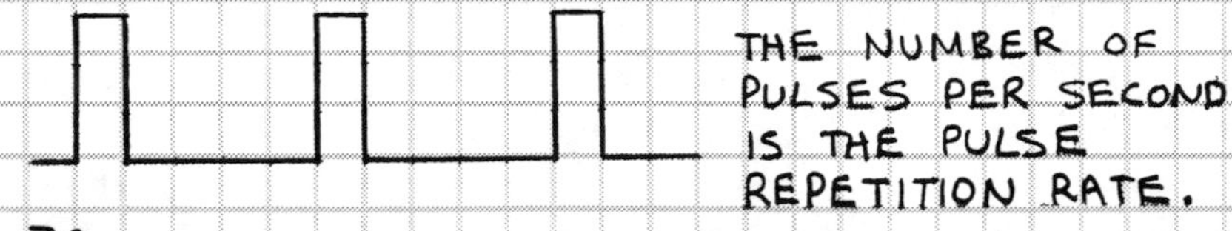

THE NUMBER OF PULSES PER SECOND IS THE PULSE REPETITION RATE.

# SIGNALS

ELECTRONIC SIGNALS RANGE FROM AUDIBLE TONES TO COMPLEX INFORMATION CARRIED BY A FLUCTUATING (ANALOG) OR PULSATING (DIGITAL) WAVE, CURRENT OR VOLTAGE. MANY MODULATION METHODS ARE USED TO IMPRESS A SIGNAL ON A CARRIER.

## MODULATION METHODS

### ANALOG

UNMODULATED CARRIER WAVE

ANALOG SIGNAL

AMPLITUDE MODULATION

FREQUENCY MODULATION

### PULSE

ANALOG SIGNAL

PULSE AMPLITUDE

PULSE DURATION

PULSE FREQUENCY

### DIGITAL

BINARY BIT PATTERN 0 0 0 1 0 1 0 1 1 0 0

NON-RETURN TO ZERO (NRZ)

RETURN TO ZERO (RZ)

MANCHESTER

FREQUENCY SHIFT KEYING (FSK)

# 4. CODES AND SYMBOLS

## ALPHABET, ASCII & MORSE CODE

| ALPHABET | ASCII | MORSE CODE |
|---|---|---|
| A | 100 0001 | .- |
| B | 100 0010 | -... |
| C | 100 0011 | -.-. |
| D | 100 0100 | -.. |
| E | 100 0101 | . |
| F | 100 0110 | ..-. |
| G | 100 0111 | --. |
| H | 100 1000 | .... |
| I | 100 1001 | |
| J | 100 1010 | .--- |
| K | 100 1011 | -.- |
| L | 100 1100 | .-.. |
| M | 100 1101 | -- |
| N | 100 1110 | -. |
| O | 100 1111 | --- |
| P | 101 0000 | .--. |
| Q | 101 0001 | --.- |
| R | 101 0010 | .-. |
| S | 101 0011 | ... |
| T | 101 0100 | - |
| U | 101 0101 | ..- |
| V | 101 0110 | ...- |
| W | 101 0111 | .-- |
| X | 101 1000 | -..- |
| Y | 101 1001 | -.-- |
| Z | 101 1010 | --.. |
| 0 | 011 0000 | ----- |
| 1 | 011 0001 | .---- |
| 2 | 011 0010 | ..--- |
| 3 | 011 0011 | ...-- |
| 4 | 011 0100 | ....- |
| 5 | 011 0101 | ..... |
| 6 | 011 0110 | -.... |
| 7 | 011 0111 | --... |
| 8 | 011 1000 | ---.. |
| 9 | 011 1001 | ----. |

# ASCII

| | | | | | | | | 0 | 0 | 1 | 1 | 1 | 1 |
|---|---|---|---|---|---|---|---|---|---|---|---|---|---|
| | | | | | | | | 1 | 1 | 0 | 0 | 1 | 1 |
| | | | | | | | | 0 | 1 | 0 | 1 | 0 | 1 |
| | | | | | | | COLUMN → / ROW ↓ | 0 & 1 | 2 | 3 | 4 | 5 | 6 | 7 |
| ↓ | ↓ | ↓ | 0 | 0 | 0 | 0 | 0 | | SP | 0 | @ | P | ` | p |
| | | | 0 | 0 | 0 | 1 | 1 | | ! | 1 | A | Q | a | q |
| | | | 0 | 0 | 1 | 0 | 2 | | " | 2 | B | R | b | r |
| | | | 0 | 0 | 1 | 1 | 3 | | # | 3 | C | S | c | s |
| | | | 0 | 1 | 0 | 0 | 4 | | $ | 4 | D | T | d | t |
| | | | 0 | 1 | 0 | 1 | 5 | | % | 5 | E | U | e | u |
| | | | 0 | 1 | 1 | 0 | 6 | | & | 6 | F | V | f | v |
| | | | 0 | 1 | 1 | 1 | 7 | | ' | 7 | G | W | g | w |
| | | | 1 | 0 | 0 | 0 | 8 | | ( | 8 | H | X | h | x |
| | | | 1 | 0 | 0 | 1 | 9 | | ) | 9 | I | Y | i | y |
| | | | 1 | 0 | 1 | 0 | 10 | | * | : | J | Z | j | z |
| | | | 1 | 0 | 1 | 1 | 11 | | + | ; | K | [ | k | { |
| | | | 1 | 1 | 0 | 0 | 12 | | , | < | L | \ | l | ¦ |
| | | | 1 | 1 | 0 | 1 | 13 | | — | = | M | ] | m | } |
| | | | 1 | 1 | 1 | 0 | 14 | | . | > | N | ^ | n | ~ |
| | | | 1 | 1 | 1 | 1 | 15 | | / | ? | O | _ | o | DEL |

SP - SPACE

CONTROL CHARACTERS (NON PRINTING) ↑ (column 0 & 1)

ASCII - AMERICAN STANDARD CODE FOR INFORMATION INTERCHANGE. ASCII IS THE PRINCIPLE COMPUTER KEYBOARD CODE. ASSEMBLY LANGUAGE PROGRAMMERS CONVERT BINARY ASCII (ABOVE) TO HEXADECIMAL. PRINCIPLE HEX EQUIVALENTS:

| | | | | | |
|---|---|---|---|---|---|
| A- 41 | G- 47 | M- 4D | S- 53 | Y- 59 | 4- 34 |
| B- 42 | H- 48 | N- 4E | T- 54 | Z- 5A | 5- 35 |
| C- 43 | I- 49 | O- 4F | U- 55 | Ø- 30 | 6- 36 |
| D- 44 | J- 4A | P- 50 | V- 56 | 1- 31 | 7- 37 |
| E- 45 | K- 4B | Q- 51 | W- 57 | 2- 32 | 8- 38 |
| F- 46 | L- 4C | R- 52 | X- 58 | 3- 33 | 9- 39 |

# GREEK ALPHABET

| NAME | U | L | NAME | U | L |
|---|---|---|---|---|---|
| ALPHA | Α | α | NU | Ν | ν |
| BETA | Β | β | XI | Ξ | ξ |
| GAMMA | Γ | γ | OMICRON | Ο | ο |
| DELTA | Δ | δ | PI | Π | π |
| EPSILON | Ε | ϵ | RHO | Ρ | ρ |
| ZETA | Ζ | ζ | SIGMA | Σ | σ |
| ETA | Η | η | TAU | Τ | τ |
| THETA | Θ | θ | UPSILON | Υ | υ |
| IOTA | Ι | ι | PHI | Φ | φ |
| KAPPA | Κ | κ | CHI | Χ | χ |
| LAMBDA | Λ | λ | PSI | Ψ | ψ |
| MU | Μ | μ | OMEGA | Ω | ω |

U–UPPER CASE L–LOWER CASE

# COMMON GREEK SYMBOLS

| LETTER | SYMBOLIZES OR DESIGNATES |
|---|---|
| α | ANGLES, ACCELERATION, AREA |
| β | ANGLES, |
| γ | CONDUCTIVITY, SPECIFIC GRAVITY |
| Δ | INCREMENT, DECREMENT |
| ϵ | DIELECTRIC CONSTANT |
| Ε | ENERGY |
| Ζ | IMPEDANCE |
| η | FM MODULATION INDEX |
| θ | ANGLES, TIME CONSTANT, TEMPERATURE |
| λ | WAVELENGTH, CONDUCTIVITY |
| μ | MICRO (PREFIX), AMPLIFICATION FACTOR |
| ν | FREQUENCY |
| π | CIRCUMFERENCE ÷ DIAMETER (3.14159...) |
| ρ | RESISTIVITY, REFLECTANCE |
| Σ | SUMMATION SIGN |
| τ | TIME CONSTANT, TRANSMITTANCE |
| φ | ANGLE, RADIANT POWER |
| ω | ANGLE, ANGULAR FREQUENCY |
| Ω | SOLID ANGLE, RESISTANCE (OHMS) |

# RESISTOR COLOR CODE

| COLOR | SIGNIFICANT DIGITS (1 & 2) | MULTIPLIER (3) | TOL (4) |
|---|---|---|---|
| BLACK | 0 | 1 | |
| BROWN | 1 | 10 | ± 1% |
| RED | 2 | 100 | |
| ORANGE | 3 | 1,000 | |
| YELLOW | 4 | 10,000 | NO COLOR BAND: ± 20% |
| GREEN | 5 | 100,000 | |
| BLUE | 6 | 1,000,000 | |
| VIOLET | 7 | 10,000,000 | |
| GRAY | 8 | 100,000,000 | |
| WHITE | 9 | — | |
| GOLD | — | — | ± 5% |
| SILVER | — | — | ± 10% |

EXAMPLE:

1 2 3 4

1 = BROWN = 1
2 = BLACK = 0
3 = YELLOW = × 10,000
4 = SILVER = ± 10% TOLERANCE

100,000 Ω
± 10%

# TRANSFORMER COLOR CODE

AUDIO INTERSTAGE AND OUTPUT:

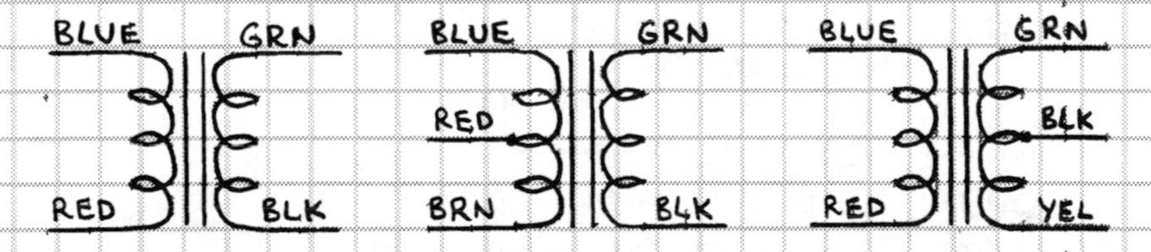


POWER: UNTAPPED PRIMARY – BLACK; FILAMENT SECONDARY – GREEN (ADDITIONAL FILAMENT – YELLOW, BROWN AND SLATE); HIGH-VOLTAGE SECONDARY – RED. COLORS MAY VARY.

NOTE: THESE ARE EIA RECOMMENDED COLORS. SEE TRANSFORMER SPECIFICATIONS TO VERIFY CODE.

# 5. ELECTRONIC ABBREVIATIONS

AC - ALTERNATING CURRENT
AF - AUDIO FREQUENCY
AFC - AUTOMATIC FREQUENCY CONTROL
AGC - AUTOMATIC GAIN CONTROL
AM - AMPLITUDE MODULATION
AMP - AMPLIFIER
ANL - AUTOMATIC NOISE LIMITER
ANT - ANTENNA
AVC - AUTOMATIC VOLUME CONTROL
AWG - AMERICAN WIRE GAUGE
B - BASE OF TRANSISTOR
BC - BROADCAST
BFO - BEAT FREQUENCY OSCILLATOR
BP - BANDPASS
C - COLLECTOR OF TRANSISTOR
CAL - CALIBRATE
CAP - CAPACITOR
CB - CITIZENS BAND
CKT - CIRCUIT
CLK - CLOCK
CRT - CATHODE RAY TUBE
C/S - CYCLES PER SECOND (HERTZ; HZ)
CT - CENTER TAP
CW - CONTINUOUS WAVE
CY - CYCLE
°C - DEGREES CELSIUS
D - DRAIN OF FET
dB - DECIBEL
DBLR - DOUBLER
DC - DIRECT CURRENT
DEG - DEGREES
DEMOD - DEMODULATION
DF - DIRECTION FINDER
DPDT - DOUBLE POLE DOUBLE THROW
DPST - DOUBLE POLE SINGLE THROW
DSB - DOUBLE SIDEBAND
E - EMITTER OF TRANSISTOR; ENERGY
EM - ELECTROMAGNETIC
EMF - ELECTROMOTIVE FORCE
EMP - ELECTROMAGNETIC PULSE
ERP - EFFECTIVE RADIATED POWER

F - FREQUENCY
°F - DEGREES FAHRENHEIT
FDBK - FEEDBACK
FET - FIELD EFFECT TRANSISTOR
FF - FLIP FLOP
FIL - FILAMENT
FM - FREQUENCY MODULATION
FREQ - FREQUENCY
FSC - FULL SCALE
FWHM - FULL WIDTH HALF MAXIMUM
G - GATE OF FET
GA - GAUGE
GND - GROUND
HF - HIGH FREQUENCY
HI FI - HIGH FIDELITY
HV - HIGH VOLTAGE
HZ - HERTZ
I - CURRENT
IC - INTEGRATED CIRCUIT
IMPD - IMPEDANCE
IR - INFRARED
JFET - JUNCTION FIELD EFFECT TRANSISTOR
KWH - KILOWATT HOUR
LED - LIGHT EMITTING DIODE
LP - LOW PASS
LSI - LARGE SCALE INTEGRATION
MA - MILLIAMPERES
MIC - MICROPHONE
MOS - METAL-OXIDE-SEMICONDUCTOR
MOSFET - MOS FIELD EFFECT TRANSISTOR
NC - NO CONTACT
NEG - NEGATIVE
NF - NOISE FIGURE
NO - NORMALLY OPEN
NOM - NOMINAL
NPN - NEGATIVE-POSITIVE-NEGATIVE
OP AMP - OPERATIONAL AMPLIFIER
OSC - OSCILLATOR
OUT - OUTPUT
PAM - PULSE AMPLITUDE MODULATION
PC - PRINTED CIRCUIT
PCM - PULSE CODE MODULATION
PDM - PULSE DURATION MODULATION

PF - PICOFARAD
PFM - PULSE FREQUENCY MODULATION
PK - PEAK
PLL - PHASE LOCKED LOOP
PNP - POSITIVE - NEGATIVE - POSITIVE
POS - POSITIVE
POT - POTENTIOMETER
PREAMP - PREAMPLIFIER
PRI - PRIMARY
PRV - PEAK REVERSE VOLTAGE
PVC - POLYVINYL CHLORIDE
PWR - POWER
PWR SUP - POWER SUPPLY
PZ - PIEZOELECTRIC
Q - QUALITY FACTOR
QTZ - QUARTZ
R - RESISTANCE
RAD - RADIAN
RC - RESISTANCE - CAPACITANCE
RCDR - RECORDER
RCV - RECEIVE
RCVR - RECEIVER
RECHRG - RECHARGE
RECT - RECTIFIER
REF - REFERENCE
RF - RADIO FREQUENCY
RFC - RADIO FREQUENCY CHOKE
RFI - RADIO FREQUENCY INTERFERENCE
RL - RESISTANCE - INDUCTANCE
RLC - RESISTANCE - INDUCTANCE - CAPACITANCE
RLY - RELAY
RMS - ROOT MEAN SQUARE
RMT - REMOTE
ROT - ROTATE
RPM - REVOLUTIONS PER MINUTE
RPS - REVOLUTIONS PER SECOND
RTTY - RADIO TELETYPEWRITER
RY - RELAY
S - SOURCE OF FET
SB - SIDEBAND
SCR - SILICON CONTROLLED RECTIFIER
SEC - SECONDARY
SERVO - SERVOMECHANISM

SHLD - SHIELD
SIG - SIGNAL
SNR - SIGNAL-TO-NOISE RATIO (ALSO S/N)
SPDT - SINGLE POLE DOUBLE THROW
SPKR - SPEAKER
SPST - SINGLE POLE SINGLE THROW
SQ - SQUARE
SSB - SINGLE SIDEBAND
SUBMIN - SUBMINIATURE
SW - SHORTWAVE
SWL - SHORTWAVE LISTENING
SWR - STANDING WAVE RATIO
SYM - SYMBOL
T - TIME
TACH - TACHOMETER
TEL - TELEPHONE
TELECOM - TELECOMMUNICATIONS
TEMP - TEMPERATURE
TERM - TERMINAL
TRF - TUNED RADIO FREQUENCY
TTL - TRANSISTOR-TRANSISTOR LOGIC
TVI - TELEVISION INTERFERENCE
UHF - ULTRA HIGH FREQUENCY
UJT - UNIJUNCTION TRANSISTOR
UTC - COORDINATED UNIVERSAL TIME
V - VOLTAGE
VAC - VACUUM; AC VOLTAGE
VC - VOICE COIL
VCO - VOLTAGE CONTROLLED OSCILLATOR
VF - VARIABLE FREQUENCY
VHF - VERY HIGH FREQUENCY
VID - VIDEO
VLF - VERY LOW FREQUENCY
VOL - VOLUME
VOM - VOLT-OHM METER
VT - VACUUM TUBE
VOX - VOICE-OPERATED TRANSMITTER
W - WATT
WHM - WATT-HOUR METER
WV - WORKING VOLTAGE
X - REACTANCE
XMTR - TRANSMITTER
Z - IMPEDANCE

# 6. BASIC ELECTRONIC CIRCUITS

## HALF-WAVE RECTIFIER

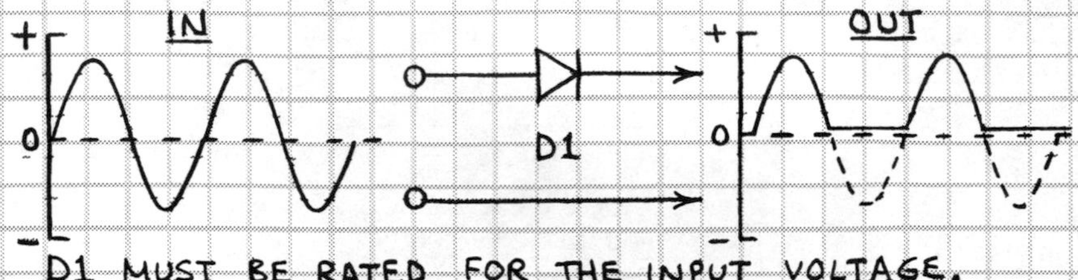


D1 MUST BE RATED FOR THE INPUT VOLTAGE.

## FULL-WAVE RECTIFIER

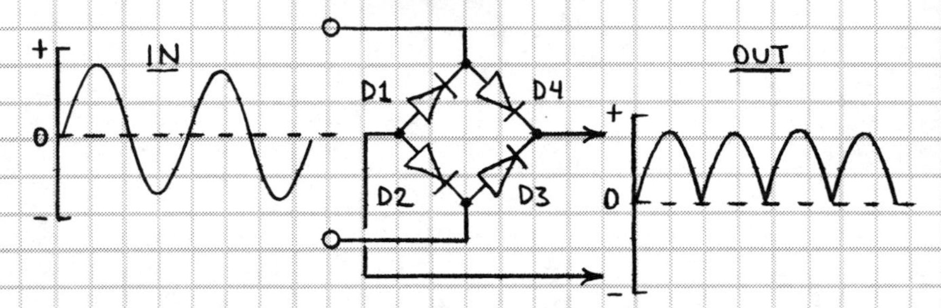


D1 – D4 MUST BE RATED FOR THE INPUT VOLTAGE.
USE INDIVIDUAL DIODES OR RECTIFIER MODULE.

## VOLTAGE DOUBLER

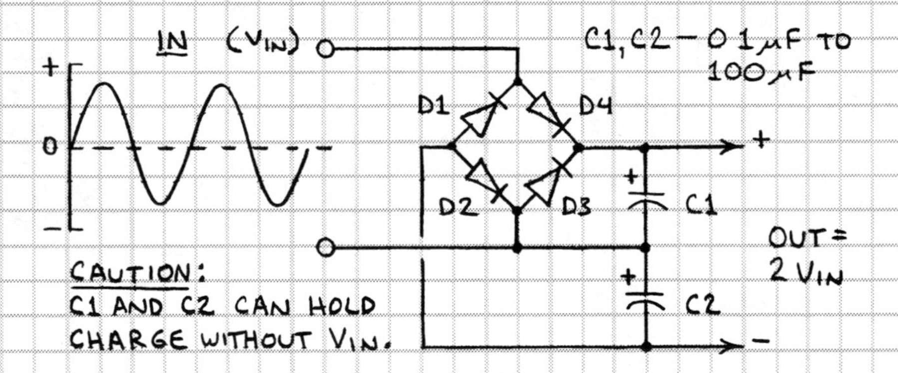


D1 – D4, C1 AND C2 MUST BE RATED FOR AT LEAST TWICE THE INPUT VOLTAGE.

# BASIC LED DRIVER

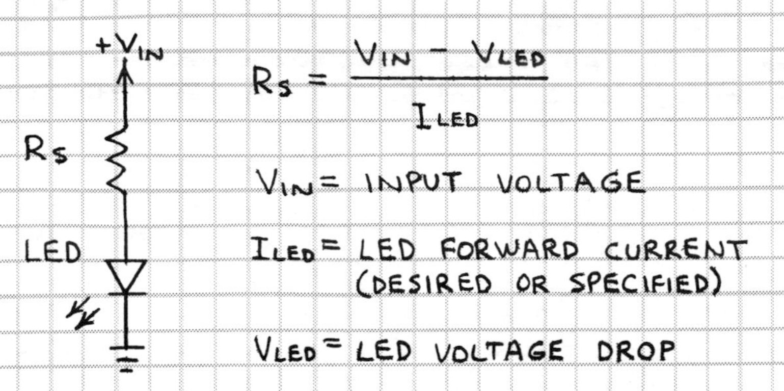


$$R_S = \frac{V_{IN} - V_{LED}}{I_{LED}}$$

$V_{IN}$ = INPUT VOLTAGE

$I_{LED}$ = LED FORWARD CURRENT (DESIRED OR SPECIFIED)

$V_{LED}$ = LED VOLTAGE DROP

EXAMPLE: ASSUME $V_{IN}$ = 9 VOLTS AND $V_{LED}$ = 1.7 VOLTS. CALCULATE VALUE OF $R_S$ FOR $I_{LED}$ = 20 mA.

$$R_S = \frac{9 - 1.7}{.02} = 365 \text{ OHMS}$$ (OK TO USE CLOSEST STANDARD VALUE)

# LOGIC GATE LED DRIVERS

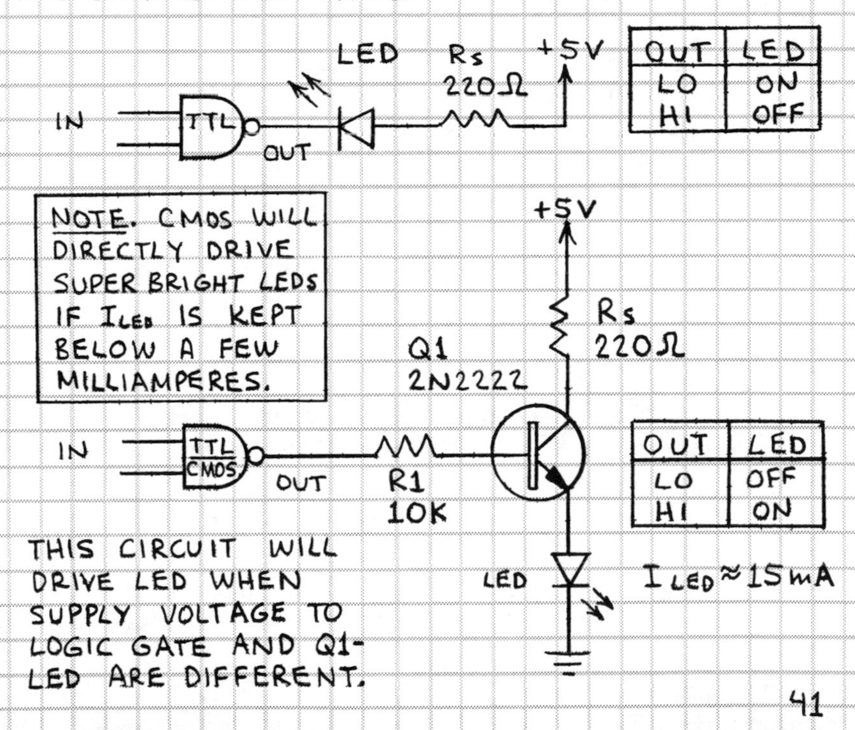


| OUT | LED |
|---|---|
| LO | ON |
| HI | OFF |

NOTE. CMOS WILL DIRECTLY DRIVE SUPER BRIGHT LEDS IF $I_{LED}$ IS KEPT BELOW A FEW MILLIAMPERES.

| OUT | LED |
|---|---|
| LO | OFF |
| HI | ON |

$I_{LED} \approx 15$ mA

THIS CIRCUIT WILL DRIVE LED WHEN SUPPLY VOLTAGE TO LOGIC GATE AND Q1-LED ARE DIFFERENT.

# INVERTING AMPLIFIER

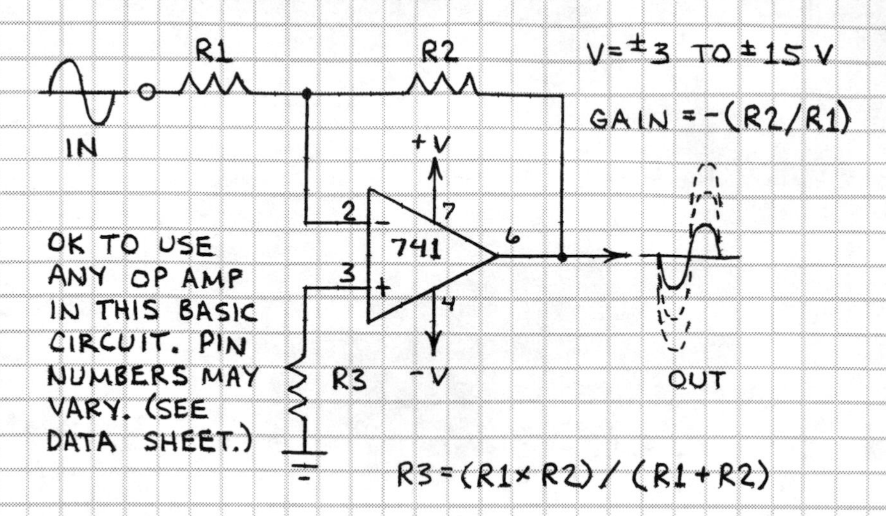


EXAMPLE: IF R1 = 4,700 OHMS AND R2 = 47,000 OHMS, THEN GAIN IS −(47,000/4,700) OR −10. R3 = 4,273 OHMS (USE CLOSEST STANDARD VALUE).

# NON-INVERTING AMPLIFIER

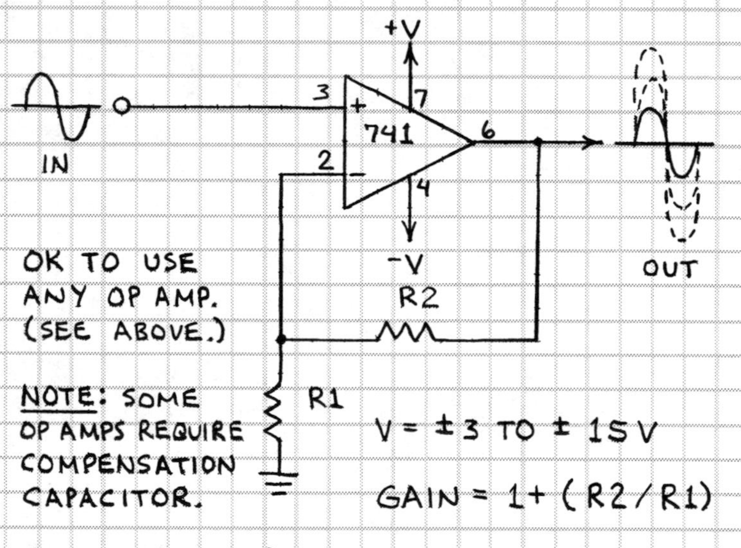


EXAMPLE: IF R1 = 4,700 OHMS AND R2 = 47,000 OHMS, THEN GAIN IS 1 + (47,000 / 4,700) OR 11.

# VOLTAGE-TO-CURRENT CONVERTER

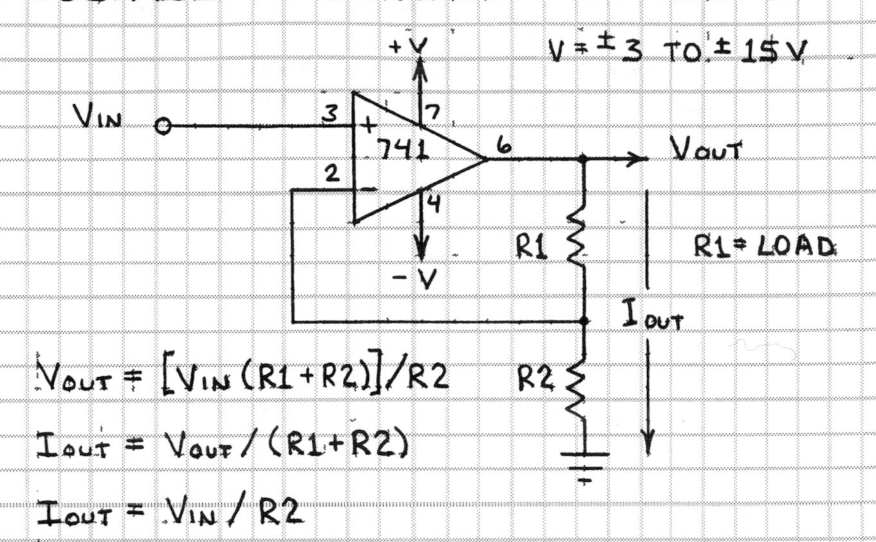


$V_{OUT} = [V_{IN}(R1+R2)]/R2$

$I_{OUT} = V_{OUT}/(R1+R2)$

$I_{OUT} = V_{IN}/R2$

EXAMPLE: ASSUME R1 IS A RESISTOR AND LED WITH COMBINED RESISTANCE OF 1,000 OHMS AND R2 IS 470 OHMS. WHEN $V_{IN}$ = 5 VOLTS, CURRENT ($I_{OUT}$) THROUGH LED IS 10.6 MA.

# CURRENT-TO-VOLTAGE CONVERTER

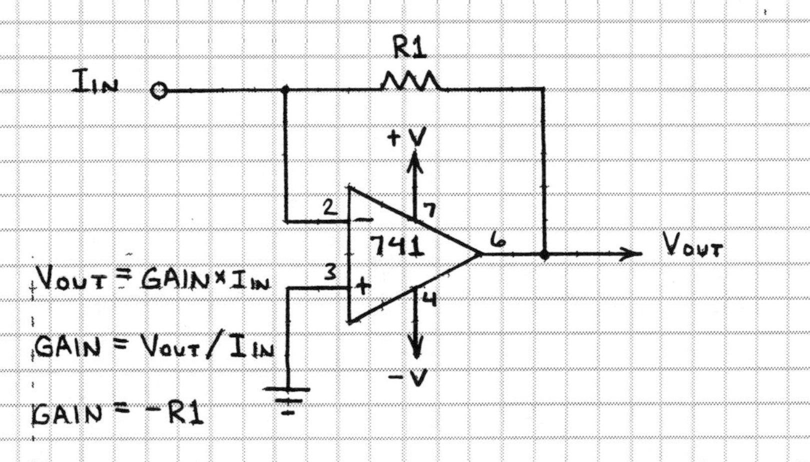


$V_{OUT} = GAIN \times I_{IN}$

$GAIN = V_{OUT}/I_{IN}$

$GAIN = -R1$

EXAMPLE: ASSUME A SOLAR CELL CONNECTED TO $I_{IN}$ DELIVERS A CURRENT OF 1 MA. IF R1 IS 1,000 OHMS, THEN $V_{OUT}$ = −(1,000 × 0.001) = −1 VOLT.

# INVERTING COMPARATOR

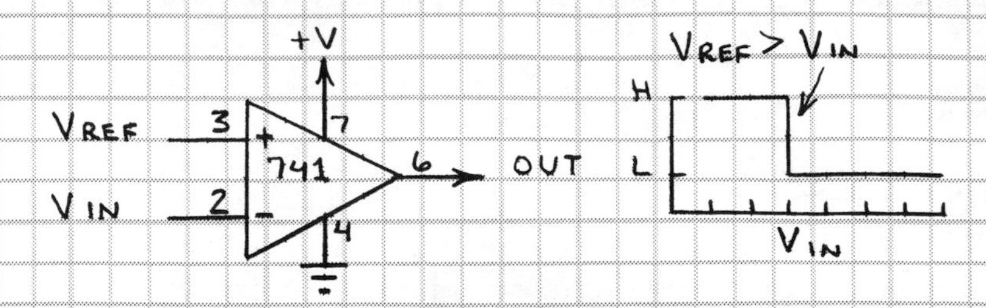


WHEN $V_{REF}$ EXCEEDS $V_{IN}$, OUTPUT SWINGS FROM HIGH TO LOW.

# NON-INVERTING COMPARATOR

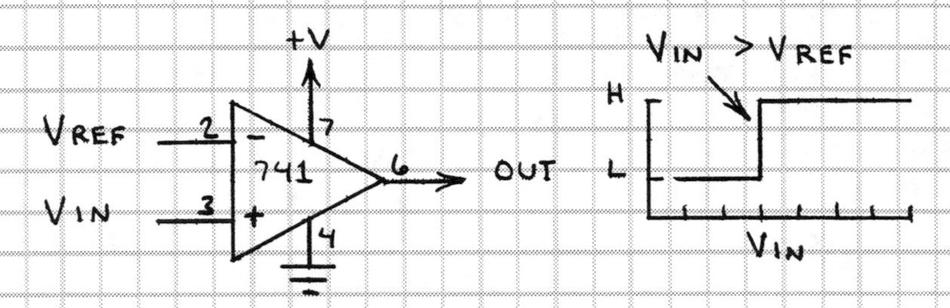


WHEN $V_{IN}$ EXCEEDS $V_{REF}$, OUTPUT SWINGS FROM LOW TO HIGH.

# WINDOW COMPARATOR

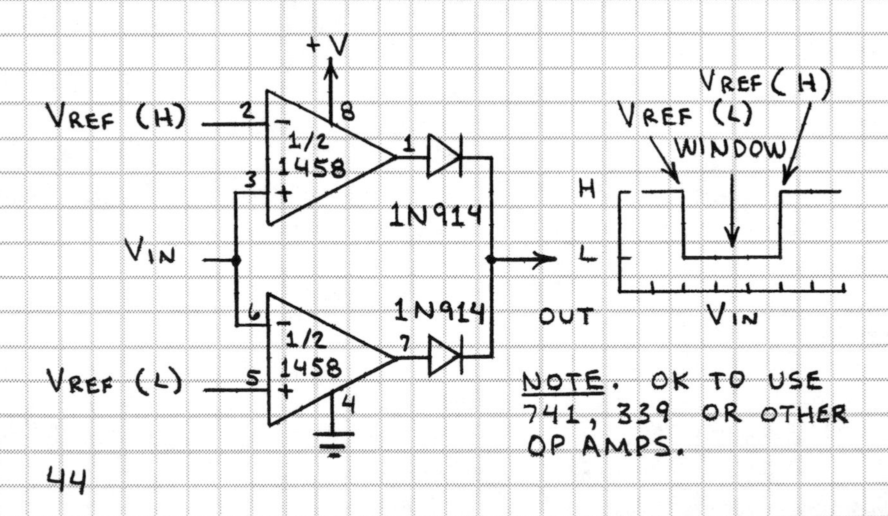


NOTE. OK TO USE 741, 339 OR OTHER OP AMPS.

# TIMER

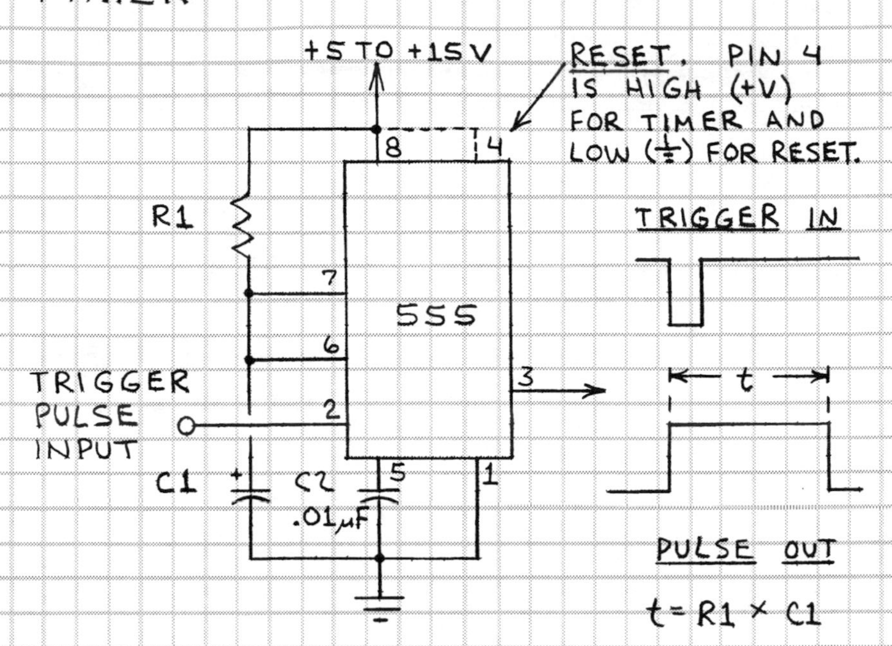


# PULSE GENERATOR

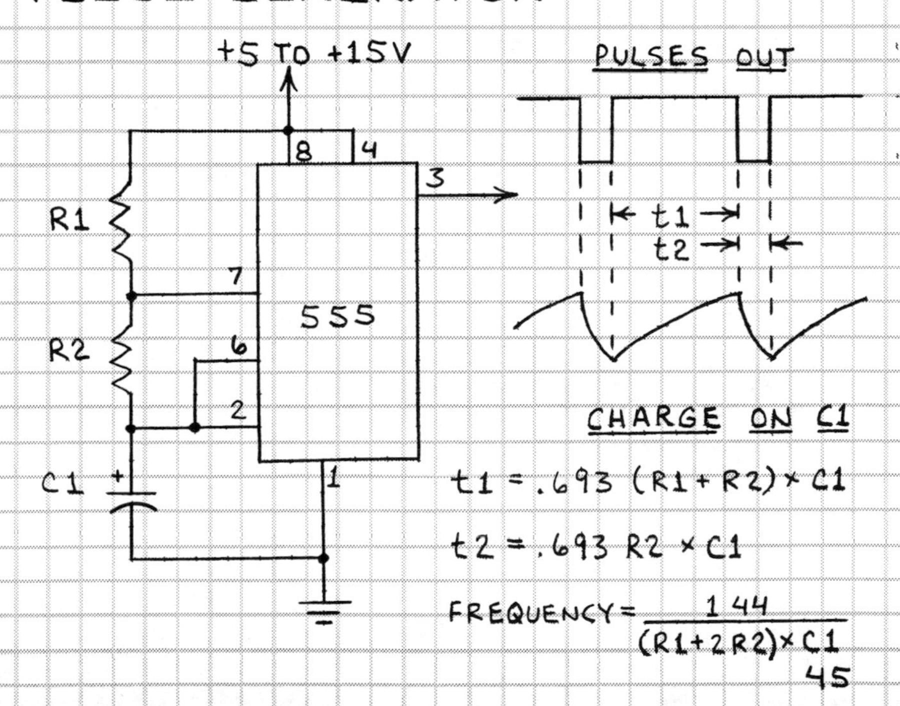

# 7. BASIC LOGIC CIRCUITS

## AND GATE

| A | B | OUT |
|---|---|---|
| L | L | L |
| L | H | L |
| H | L | L |
| H | H | H |

## NAND GATE

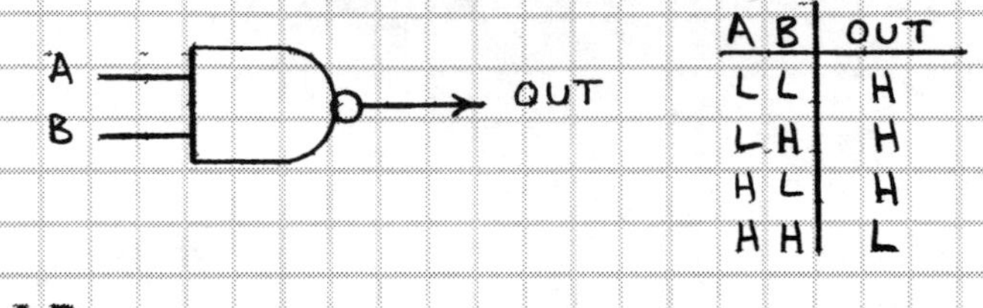


| A | B | OUT |
|---|---|---|
| L | L | H |
| L | H | H |
| H | L | H |
| H | H | L |

## OR

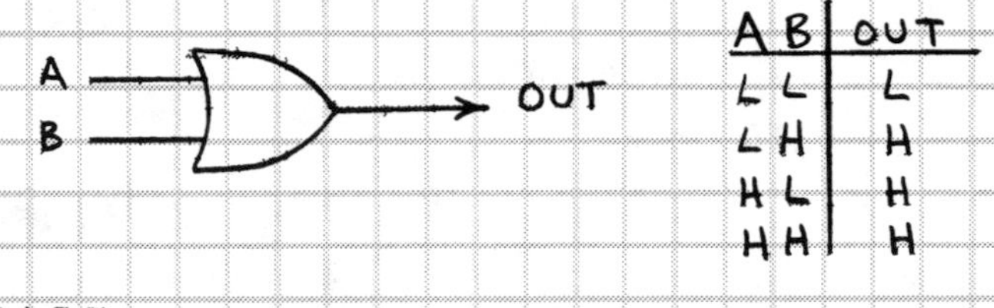


| A | B | OUT |
|---|---|---|
| L | L | L |
| L | H | H |
| H | L | H |
| H | H | H |

## NOR

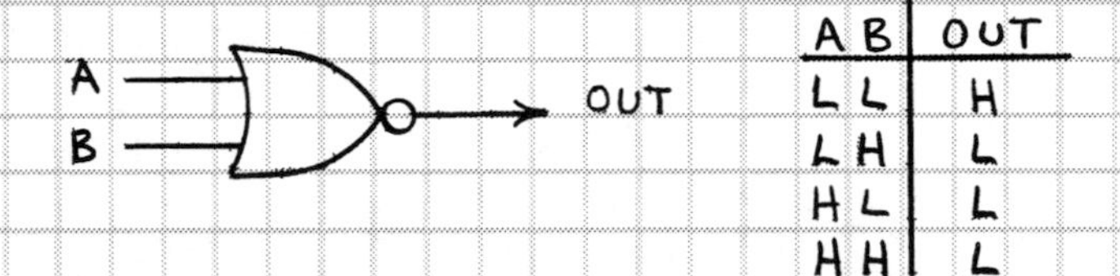


| A | B | OUT |
|---|---|---|
| L | L | H |
| L | H | L |
| H | L | L |
| H | H | L |

## EXCLUSIVE OR

| A | B | OUT |
|---|---|---|
| L | L | L |
| L | H | H |
| H | L | H |
| H | H | L |

# EXCLUSIVE NOR

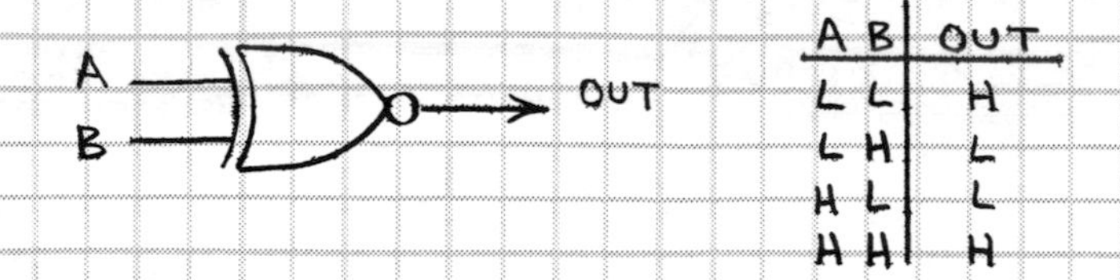


| A | B | OUT |
|---|---|---|
| L | L | H |
| L | H | L |
| H | L | L |
| H | H | H |

# BUFFER (3-STATE BUFFER)

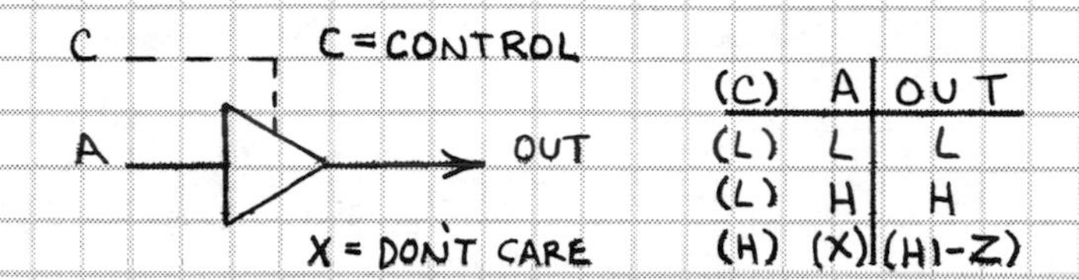


| (C) | A | OUT |
|---|---|---|
| (L) | L | L |
| (L) | H | H |
| (H) | (X) | (HI-Z) |

# INVERTER (3-STATE INVERTER)

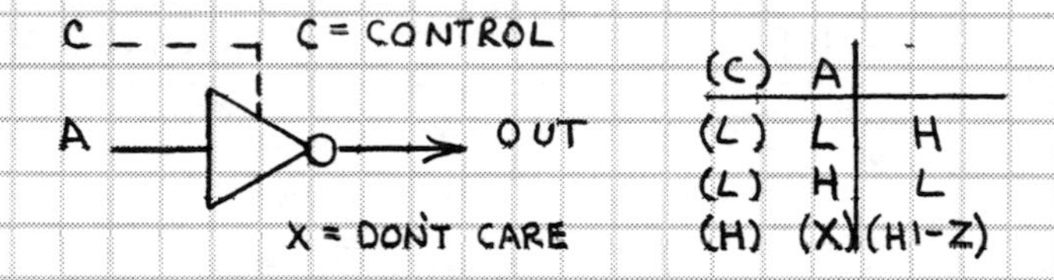


| (C) | A | |
|---|---|---|
| (L) | L | H |
| (L) | H | L |
| (H) | (X) | (HI-Z) |

# 3-STATE BUS

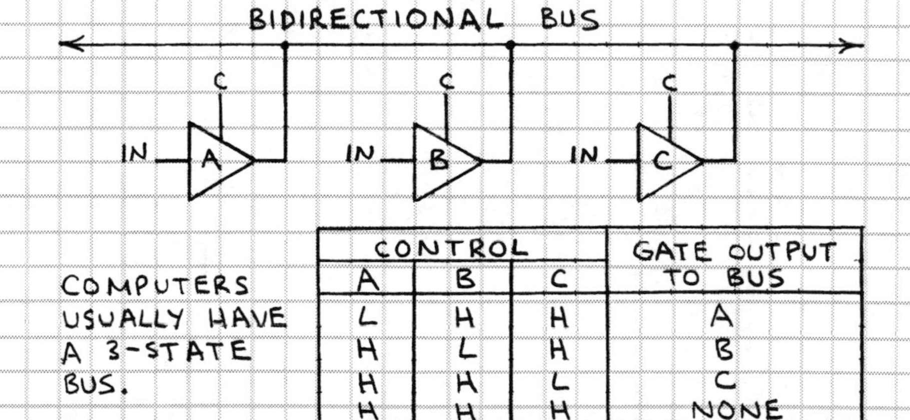


| CONTROL | | | GATE OUTPUT TO BUS |
|---|---|---|---|
| A | B | C | |
| L | H | H | A |
| H | L | H | B |
| H | H | L | C |
| H | H | H | NONE |

COMPUTERS USUALLY HAVE A 3-STATE BUS.

# RS FLIP-FLOP (LATCH)

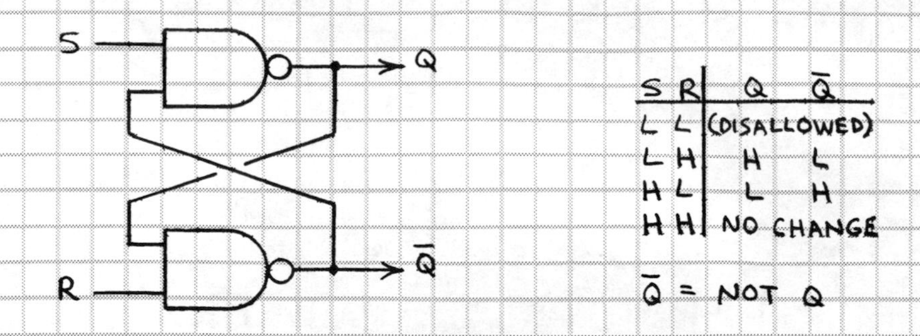


| S | R | Q | $\bar{Q}$ |
|---|---|---|---|
| L | L | (DISALLOWED) | |
| L | H | H | L |
| H | L | L | H |
| H | H | NO CHANGE | |

$\bar{Q}$ = NOT Q

# CLOCKED RS FLIP-FLOP

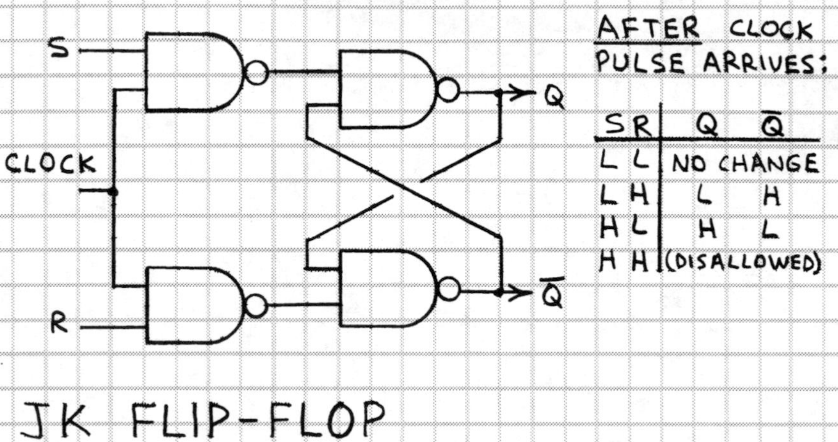


AFTER CLOCK PULSE ARRIVES:

| S | R | Q | $\bar{Q}$ |
|---|---|---|---|
| L | L | NO CHANGE | |
| L | H | L | H |
| H | L | H | L |
| H | H | (DISALLOWED) | |

# JK FLIP-FLOP

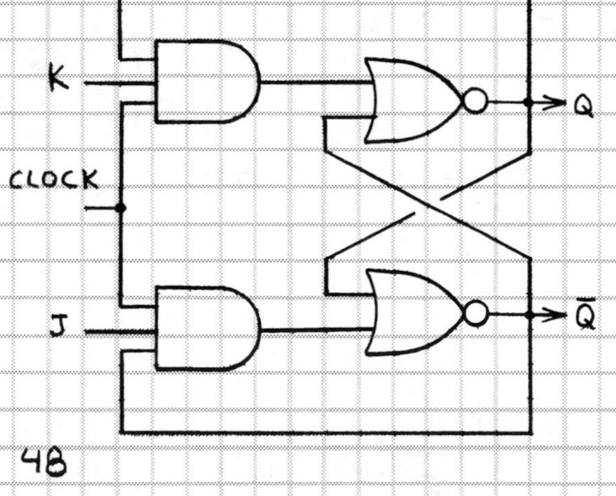


AFTER CLOCK PULSE ARRIVES'

| J | K | Q | $\bar{Q}$ |
|---|---|---|---|
| L | L | NO CHANGE | |
| L | H | L | H |
| H | L | H | L |
| H | H | TOGGLE* | |

*SEE FACING PAGE.

# D (DATA OR DELAY) FLIP-FLOP

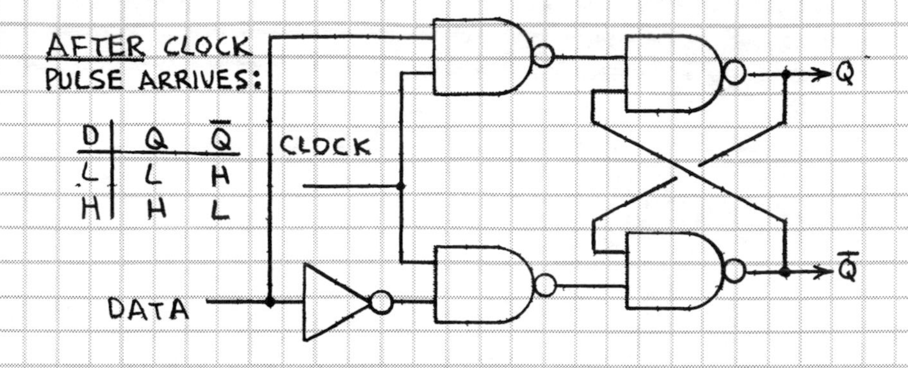


| D | Q | $\overline{Q}$ |
|---|---|---|
| L | L | H |
| H | H | L |

# T (TOGGLE) FLIP-FLOPS

THE Q (OR $\overline{Q}$) OUTPUT IS L (OR H) FOR EVERY OTHER INPUT PULSE. THEREFORE THE OUTPUT IS THE INPUT ÷ 2:

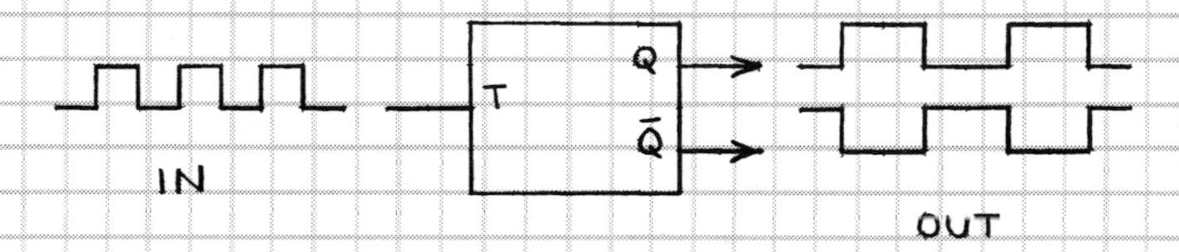


CHAINS OF T FLIP-FLOPS ARE USED TO MAKE BINARY COUNTERS. THE JK FLIP-FLOP (FACING PAGE) FUNCTIONS AS A T FLIP-FLOP WHEN BOTH THE J AND J INPUTS ARE KEPT HIGH AND INPUT PULSES ARE APPLIED TO THE CLOCK INPUT. OTHER T FLIP-FLOPS.

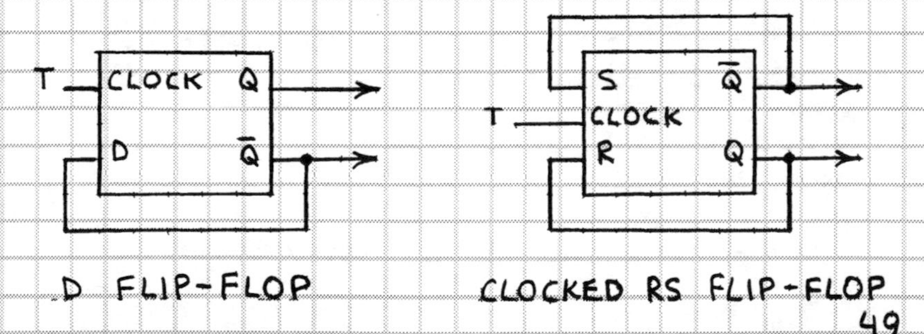


D FLIP-FLOP CLOCKED RS FLIP-FLOP

# 8. POWER SUPPLIES

## BATTERIES

### SYMBOLS

SINGLE CELL: + −   MULTIPLE CELL: + −

### CONNECTIONS

SERIES:

\+ B1 B2 −

TOTAL VOLTAGE IS SUM OF EACH CELL VOLTAGE.

PARALLEL:

\+ B1 B2 −

TOTAL CURRENT CAPACITY IS SUM OF EACH CELL CAPACITY. CELLS SHOULD HAVE EQUAL CAPACITY.

BIPOLAR:

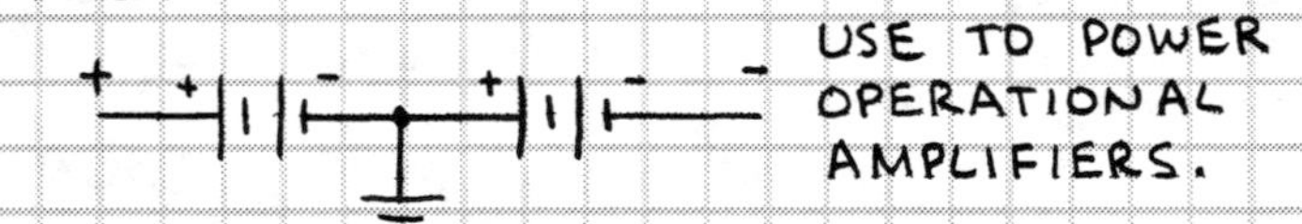

USE TO POWER OPERATIONAL AMPLIFIERS.

## STORAGE BATTERIES

STORAGE BATTERIES CAN BE USED AND RECHARGED MANY TIMES. PRINCIPLE TYPES:

LEAD-ACID — 2.0 VOLTS PER CELL. HIGH CURRENT CAPACITY. GOOD AT LOW TEMPERATURE.

NICKEL-CADMIUM (NICAD) — 1.2 VOLTS PER CELL. CAN BE STORED FOR EXTENDED TIME WHEN DISCHARGED. MANY DIFFERENT KINDS AVAILABLE. VERY ECONOMICAL POWER SOURCE.

# PRIMARY BATTERIES

PRIMARY BATTERIES ARE NOT RECHARGEABLE. CHIEF AMONG THE MANY TYPES AVAILABLE:

CARBON-ZINC — 1.5 VOLTS PER CELL. READILY AVAILABLE AND LOW COST.

ZINC-CHLORIDE — 1.5 VOLTS PER CELL. TWICE THE ENERGY DENSITY OF CARBON-ZINC.

ALKALINE — 1.5 VOLTS PER CELL. USE FOR HIGH CURRENT LOADS (MOTORS, LAMPS, ETC.).

MERCURY — 1.35 AND 1.4 VOLTS PER CELL. UNIFORM VOLTAGE DURING DISCHARGE.

SILVER OXIDE — 1.5 VOLTS PER CELL. NEARLY UNIFORM VOLTAGE DURING DISCHARGE.

LITHIUM MANGANESE — 3.0 VOLTS PER CELL. EXCEPTIONALLY LONG STORAGE LIFE. VERY HIGH ENERGY DENSITY.

# BATTERY PRECAUTIONS

1. DO NOT CHARGE PRIMARY CELLS.
2. BATTERIES MAY EXPLODE WHEN HEATED.
3. DO NOT SOLDER LEADS TO A BATTERY. USE A BATTERY CLIP OR HOLDER.
4. NEVER SHORT CIRCUIT A BATTERY'S TERMINALS.
5. MOST BATTERIES SHOULD BE REMOVED FROM EQUIPMENT IN STORAGE. EXCEPTIONS ARE STORAGE BATTERIES AND LITHIUM CELLS.
6. WHEN BATTERY LEADS EXCEED ≈ 6 INCHES, CONNECT 0.1 μF CAPACITOR ACROSS LEADS AT CIRCUIT BOARD.

# LINE-POWERED SUPPLY

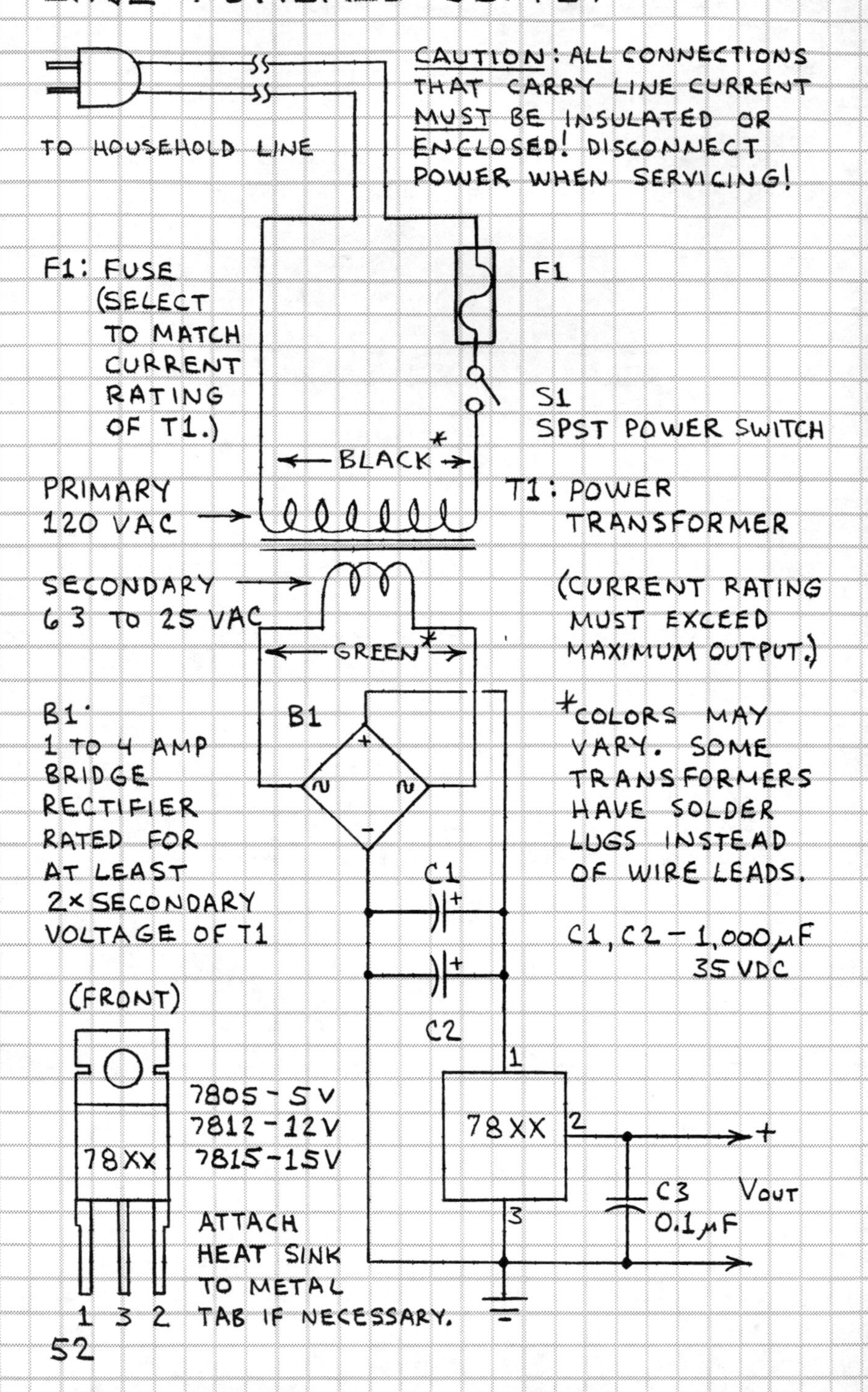

# II. SCHEMATIC SYMBOLS, DEVICE PACKAGES, DESIGN AND TESTING

## OVERVIEW

THIS SECTION COVERS ELECTRONIC PARTS AND CIRCUITS.

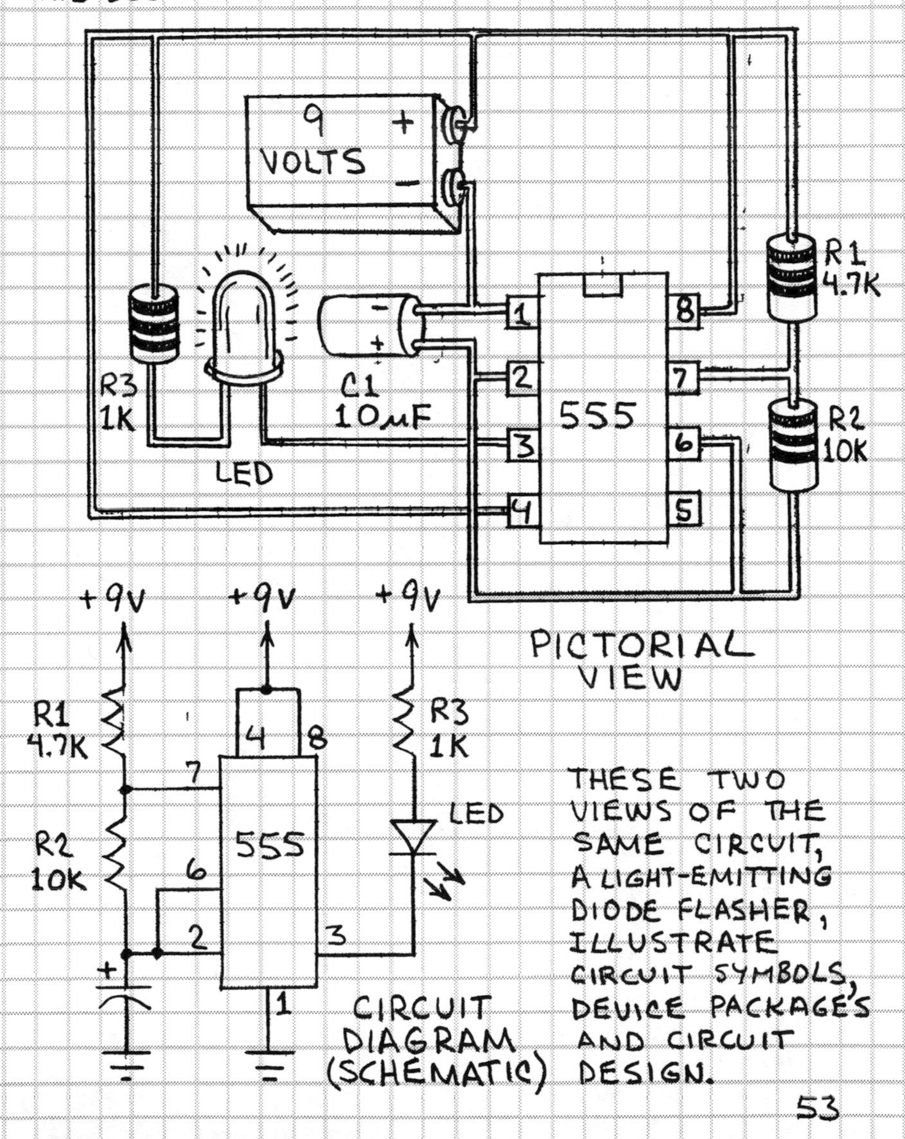


THESE TWO VIEWS OF THE SAME CIRCUIT, A LIGHT-EMITTING DIODE FLASHER, ILLUSTRATE CIRCUIT SYMBOLS, DEVICE PACKAGES AND CIRCUIT DESIGN.

# 1. SCHEMATIC SYMBOLS

## ANTENNAS

EXTERNAL

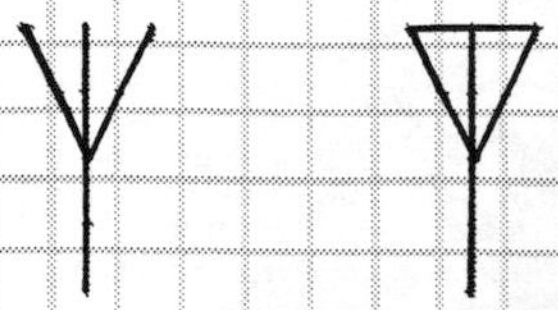

DIPOLE

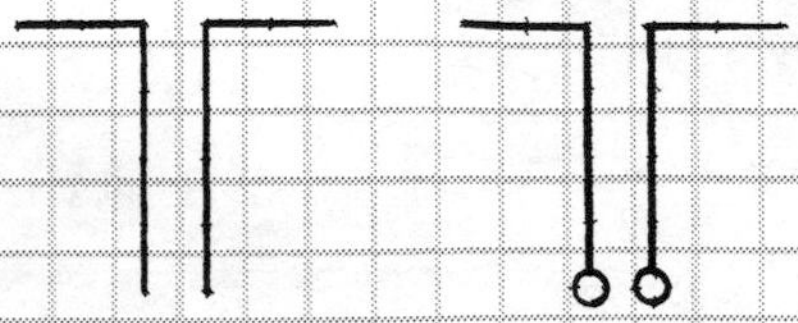

FOLDED DIPOLE

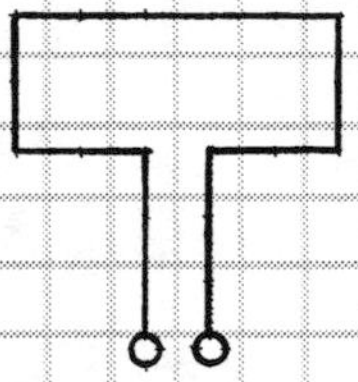

UHF LOOP

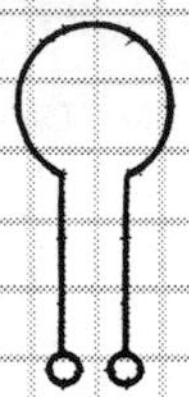

UHF BOWTIE

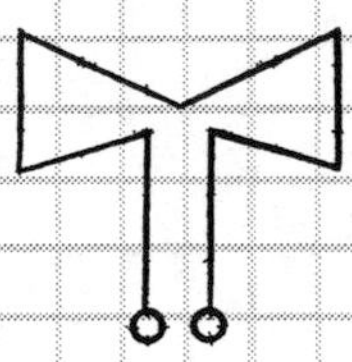

LOOP

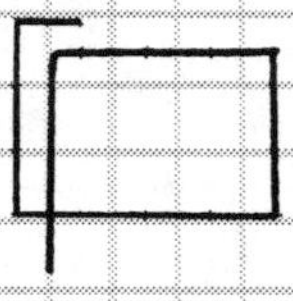

TELESCOPIC

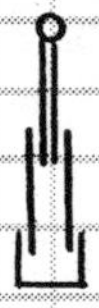

FERRITE CORE

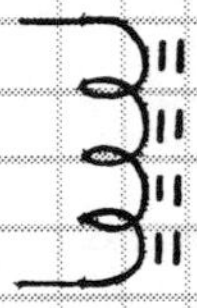

MICROWAVE HORN

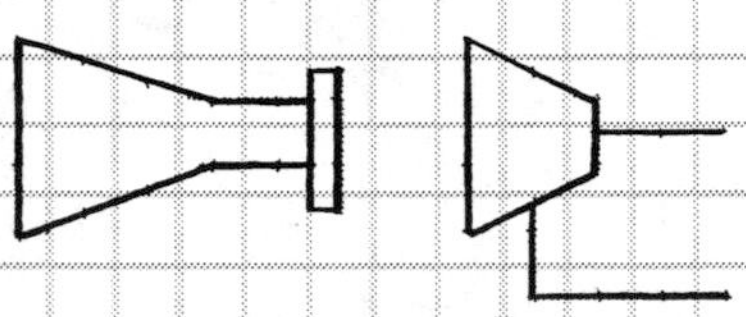

ROTATABLE LOOP

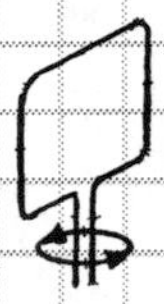

EARTH STATION

# WIRE

CONNECTED NOT CONNECTED

SHIELDED WIRE AND COAXIAL CABLE

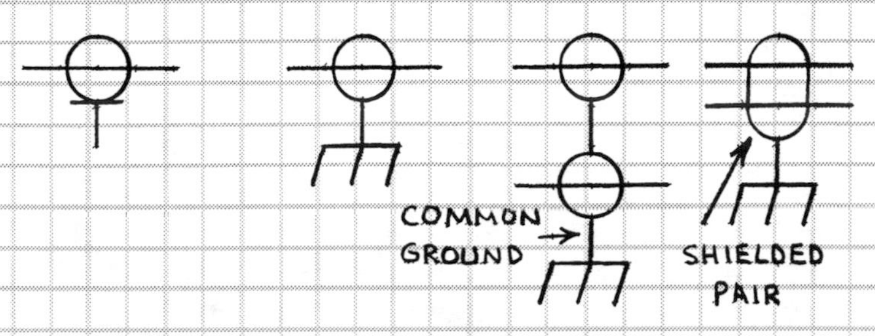


CABLE SHIELDED AT 2 POINTS

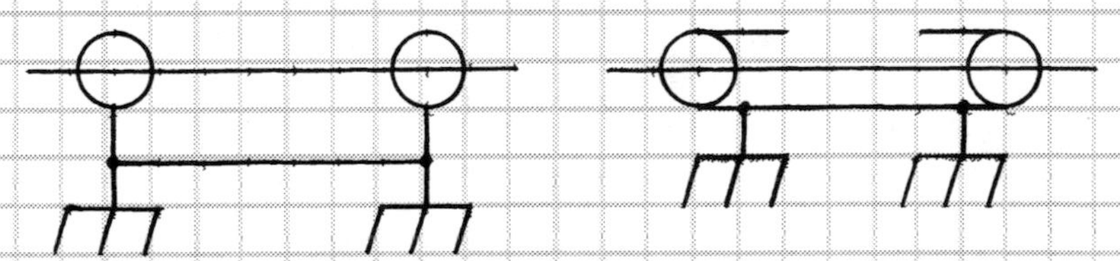

EARTH GROUND CHASSIS GROUND

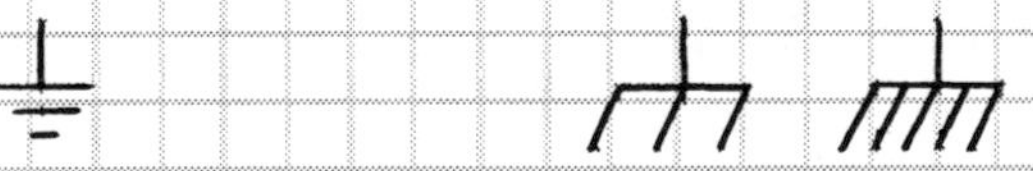

# COMMON TIE POINTS

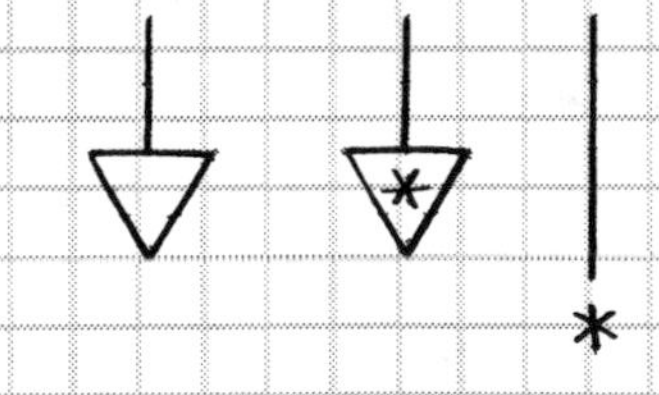

* USE FOR TWO OR MORE COMMON TIE POINTS IN SAME CIRCUIT AND INSERT NUMBER OF RELEVANT TIE POINT.

# INDUCTORS

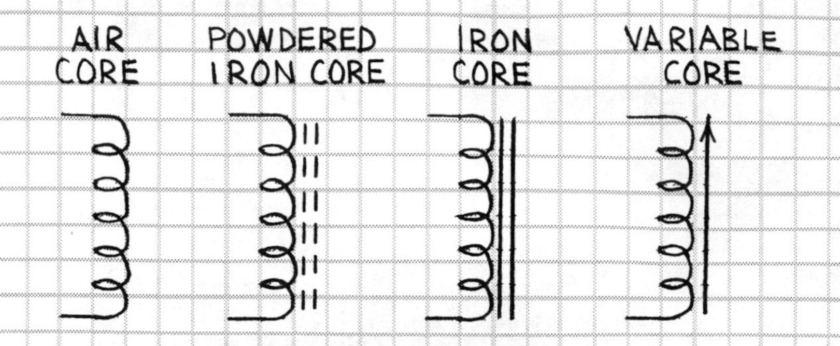


# TRANSFORMERS

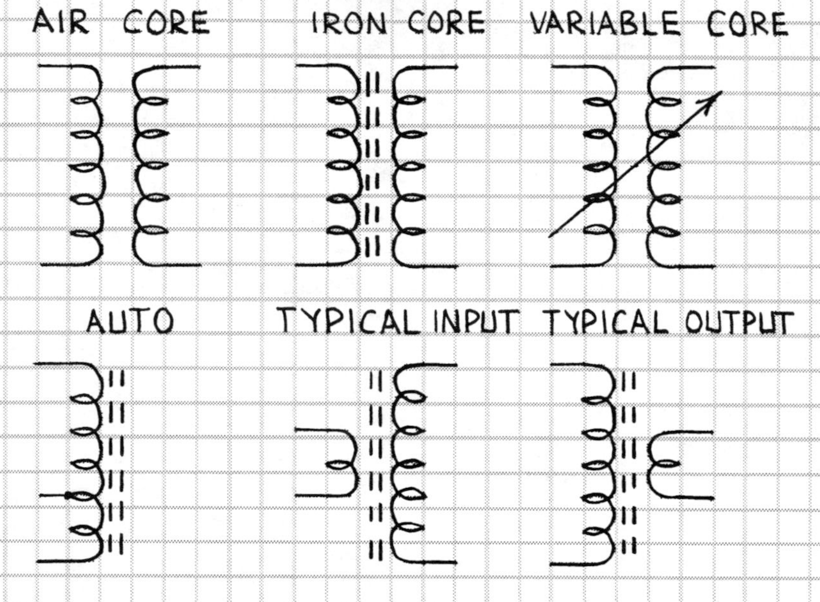


TYPICAL POWER TRANSFORMER (TAPPED)

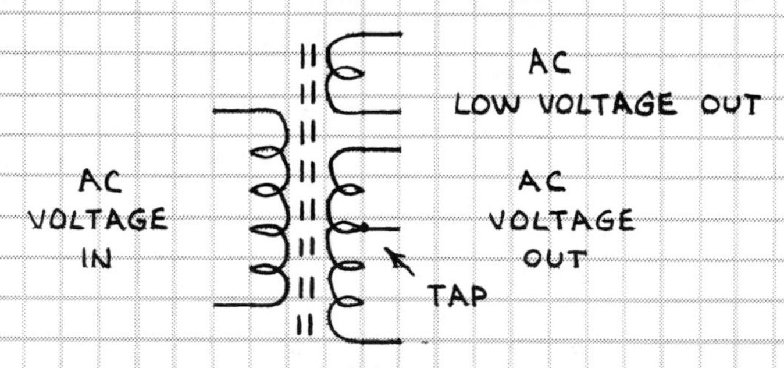

# POWER SUPPLIES

SINGLE CELL

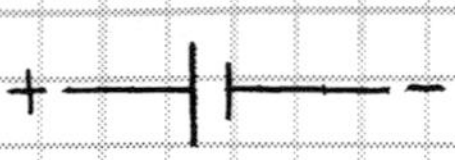

+ −

MULTIPLE CELL BATTERY

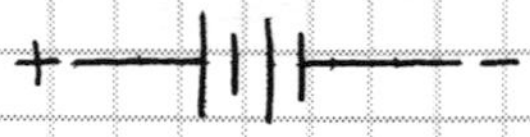

+ −

AC CURRENT SOURCES

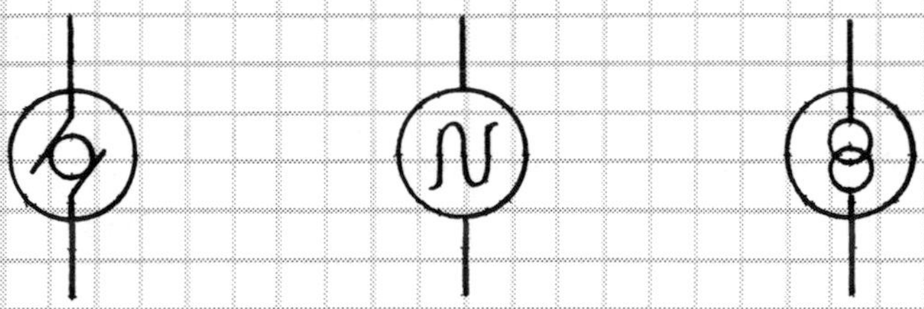

SOLAR CELLS

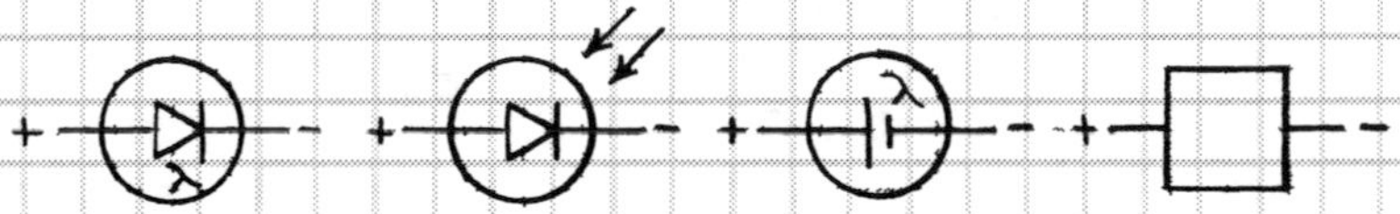

+ − + − + − + −

# FUSES

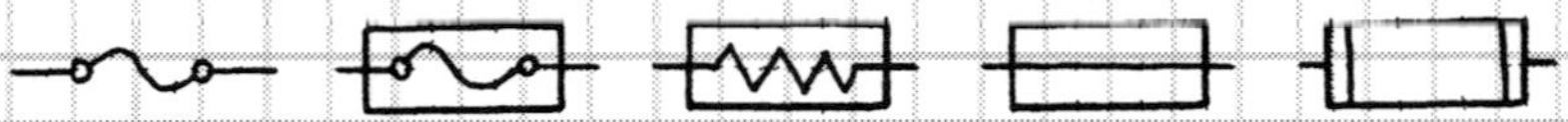

# SHIELDING

NOTE: DASHED LINE(S) ALSO USED TO INDICATE MECHANICAL CONNECTION.

# SHIELDED ENCLOSURE

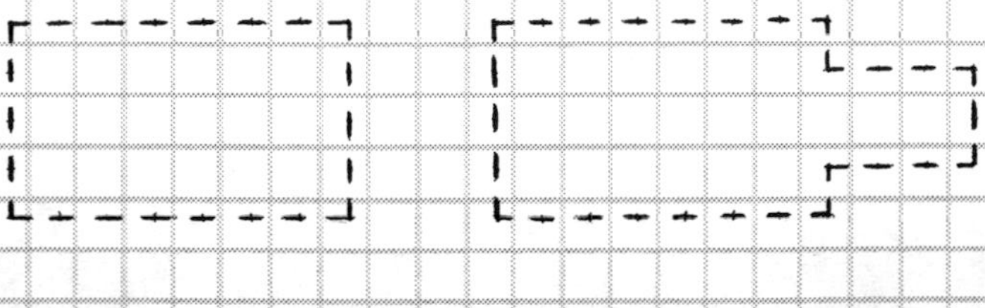

# ELECTRON TUBES

DIODE TRIODE TETRODE

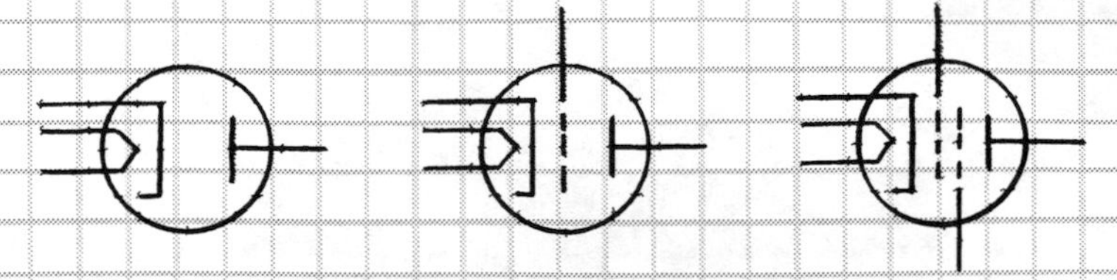

GAS-FILLED RECTIFIER FULL-WAVE RECTIFIER PHOTOTUBE

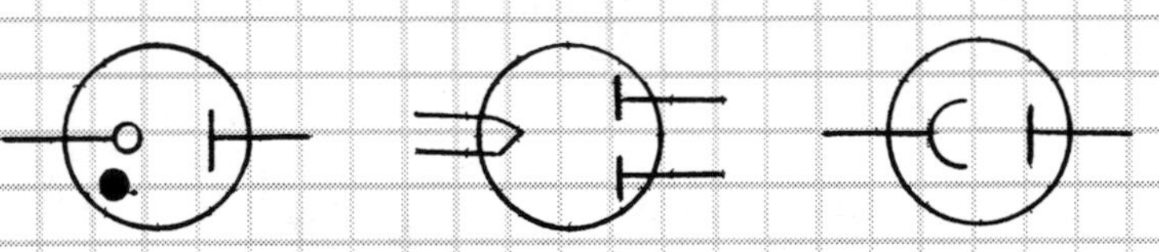

## CATHODE-RAY TUBES

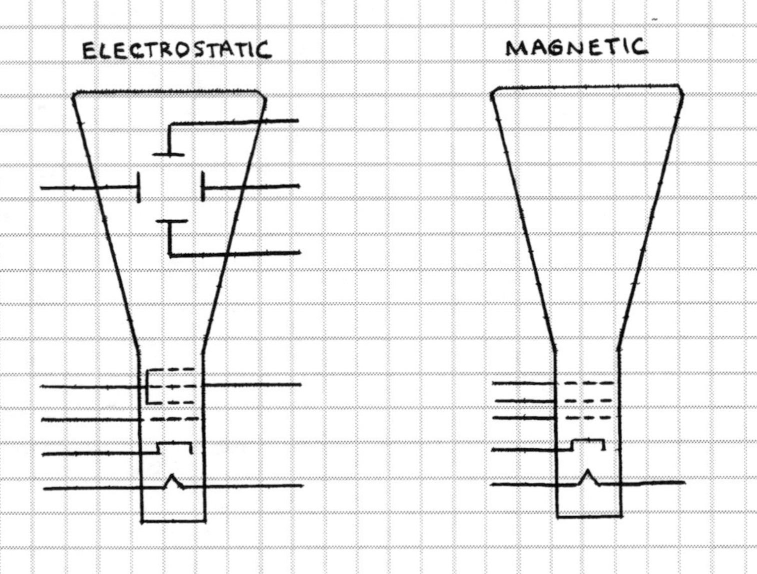


## TUBE ELEMENTS

FILAMENT CATHODE GRID PLATE

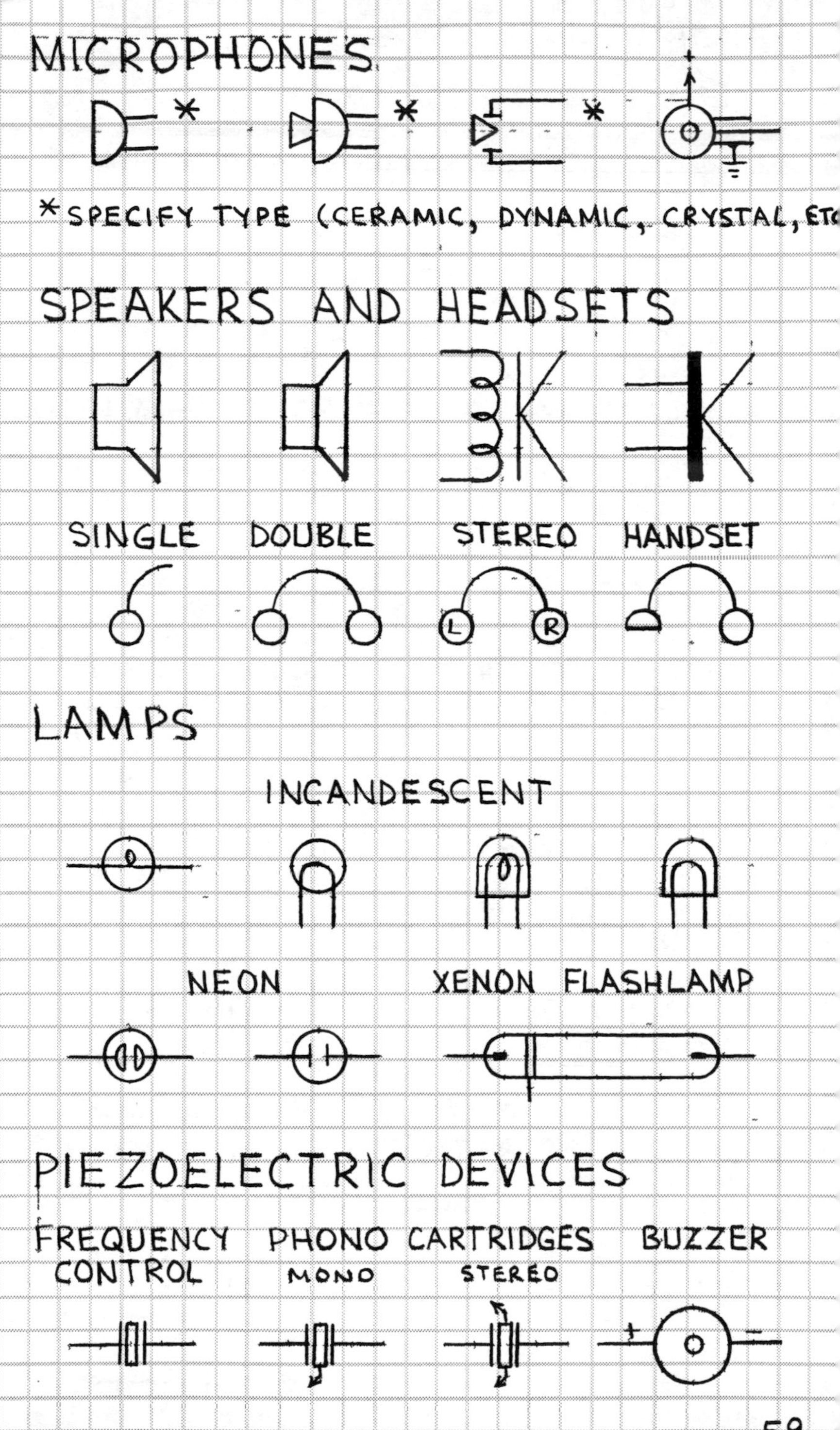

MICROPHONES
*
*
*
+
* SPECIFY TYPE (CERAMIC, DYNAMIC, CRYSTAL, ET
SPEAKERS AND HEADSETS
SINGLE
DOUBLE
STEREO
HANDSET
L
R
LAMPS
INCANDESCENT
NEON
XENON FLASHLAMP
PIEZOELECTRIC DEVICES
FREQUENCY CONTROL
PHONO CARTRIDGES
MONO
STEREO
BUZZER
+
−

# CONNECTORS

TERMINAL

TEST POINT

TP1

MALE

FEMALE

ENGAGED

PHONO/COAXIAL PLUG

PHONO/COAXIAL JACK

2-CONDUCTOR PLUG

TIP

SLEEVE

3-CONDUCTOR PLUG

TIP

RING

SLEEVE

2-CONDUCTOR JACKS

SPST SWITCH

3-CONDUCTOR JACKS

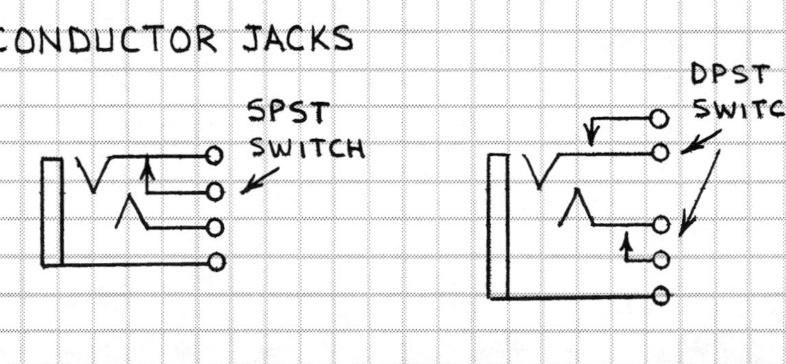

## 117-VOLT NON-POLARIZED PLUG

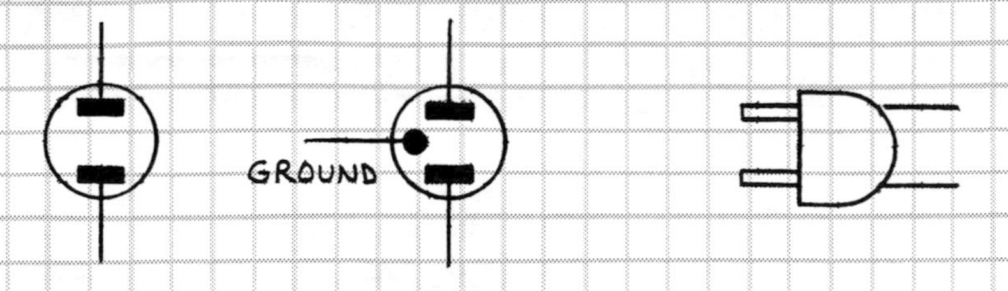


## 117-VOLT NON-POLARIZED SOCKET

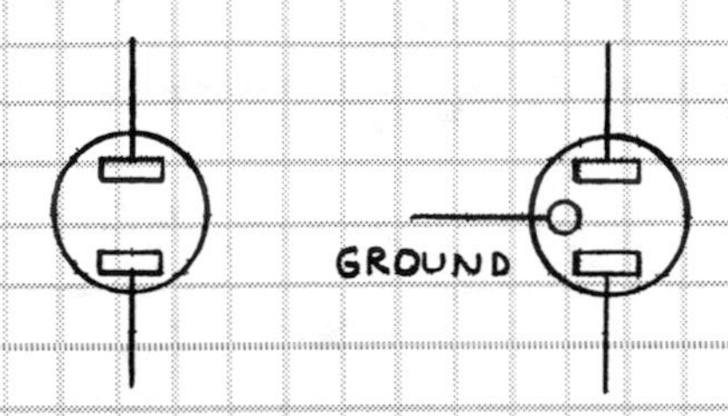


## 117-VOLT POLARIZED PLUG

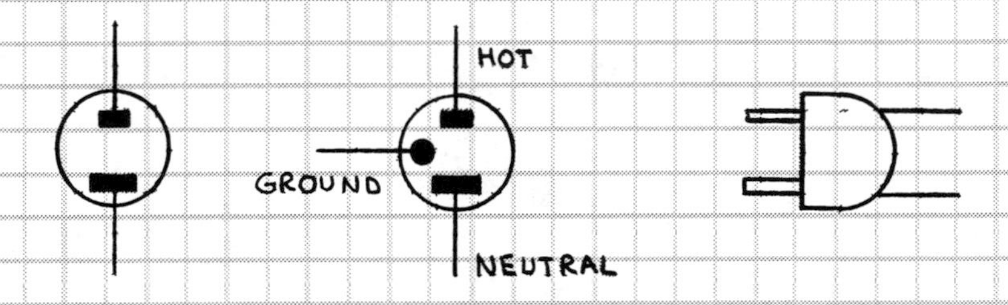


## 117-VOLT POLARIZED SOCKET

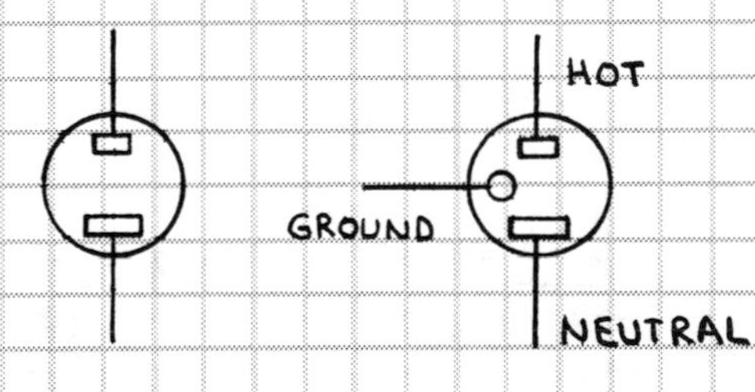


## 234-VOLT PLUG

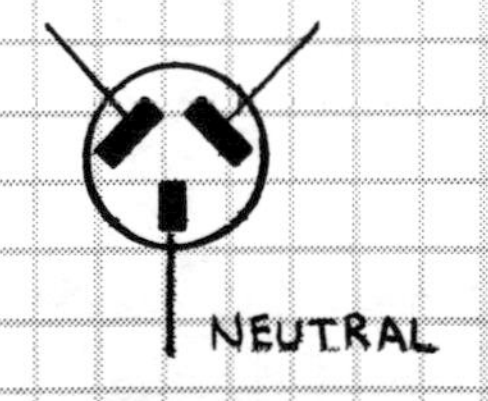


## 234-VOLT SOCKET

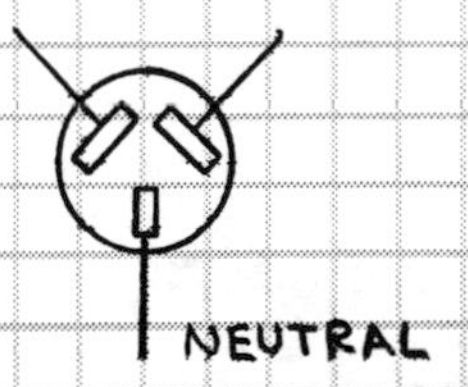

# SWITCHES

## SINGLE POLE SINGLE THROW (SPST)

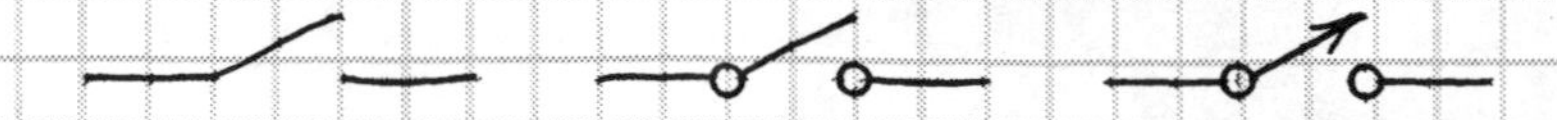

## SINGLE POLE DOUBLE THROW (SPDT)

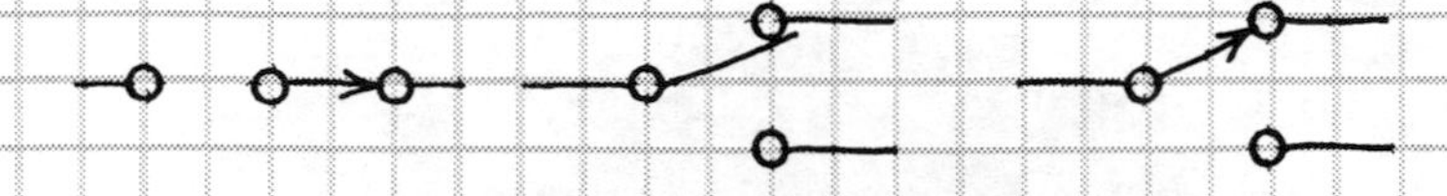

## DOUBLE POLE SINGLE THROW (DPST)

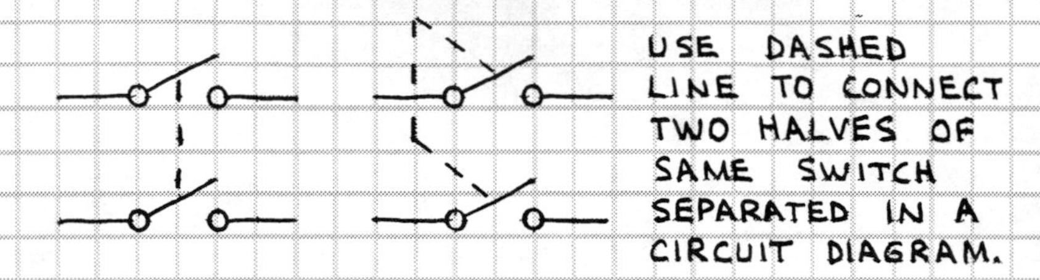


## DOUBLE POLE DOUBLE THROW (DPDT)

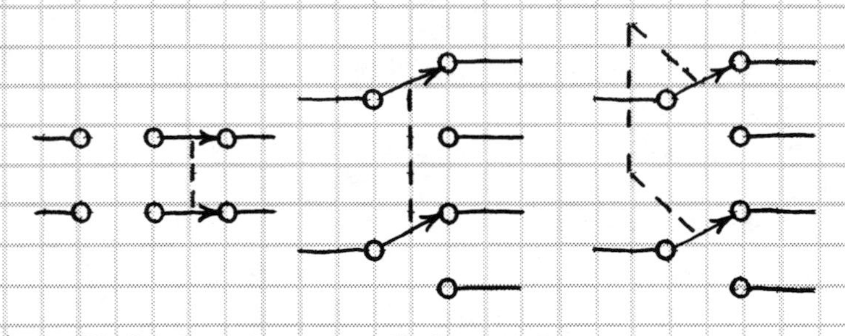

## MULTIPLE CONTACT ROTARY

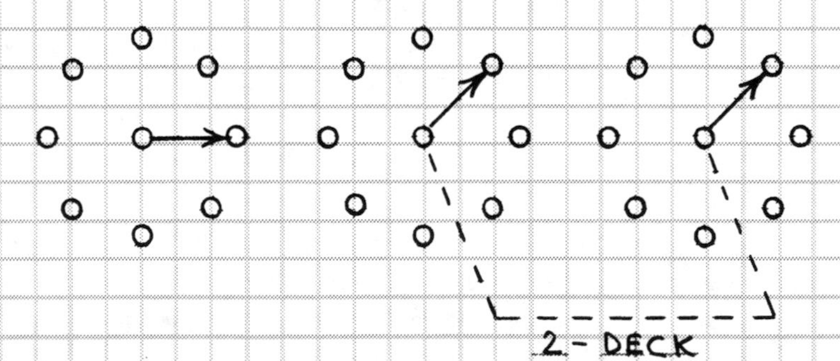

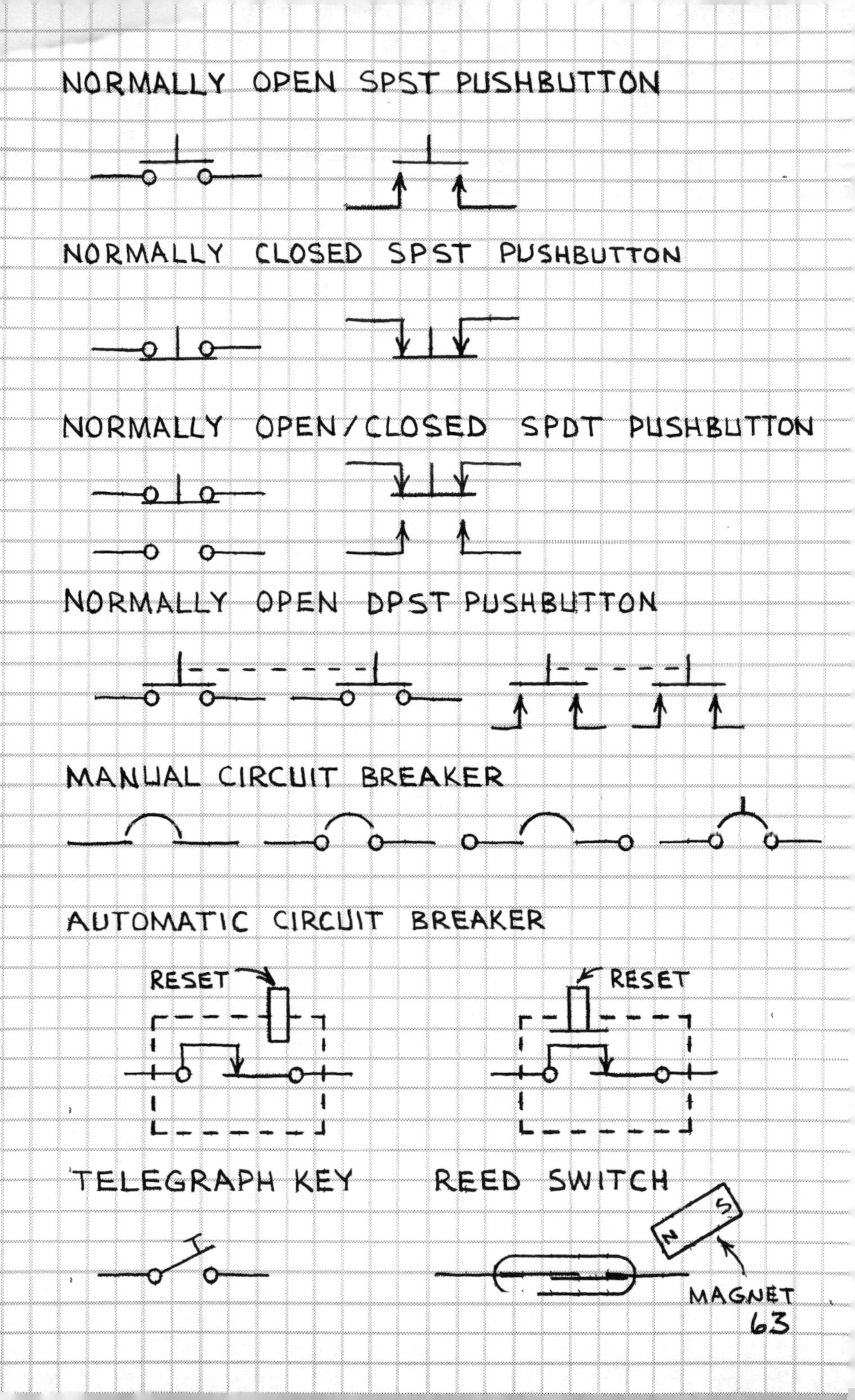
NORMALLY OPEN SPST PUSHBUTTON
NORMALLY CLOSED SPST PUSHBUTTON
NORMALLY OPEN/CLOSED SPDT PUSHBUTTON
NORMALLY OPEN DPST PUSHBUTTON
MANUAL CIRCUIT BREAKER
AUTOMATIC CIRCUIT BREAKER
RESET
RESET
TELEGRAPH KEY
REED SWITCH
N
S
MAGNET

# RELAYS

## COMPLETE RELAY SYMBOLS

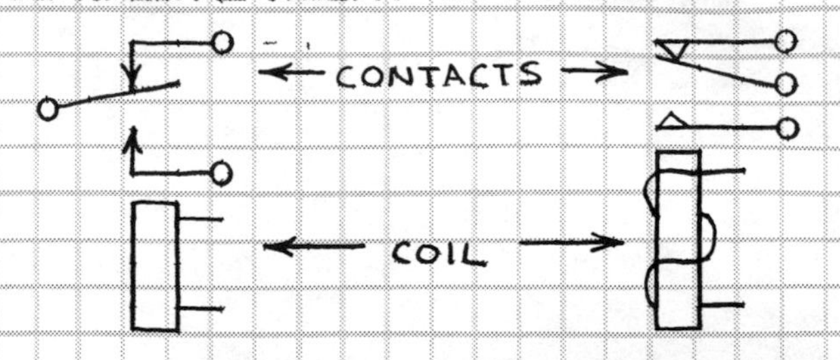


## MOST COMMON RELAY CONTACTS:

MAKE (SPST, NORMALLY OPEN)

BREAK (SPST, NORMALLY CLOSED)

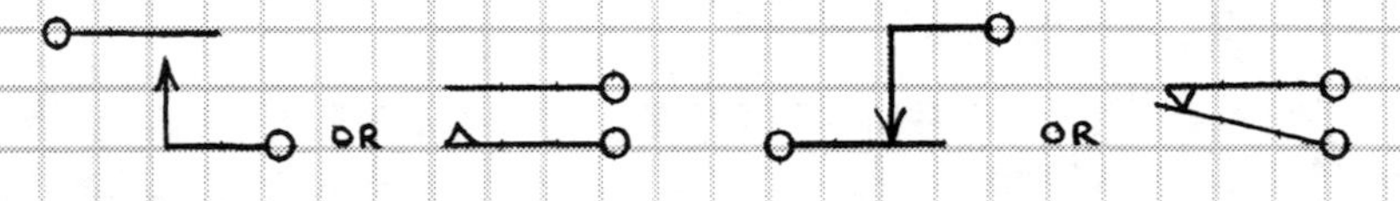


BREAK-MAKE (SPDT)

MAKE-BREAK (SPDT)

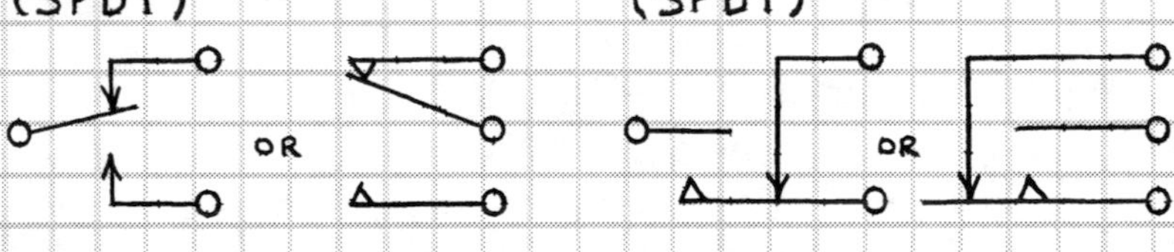


DPST

DPDT

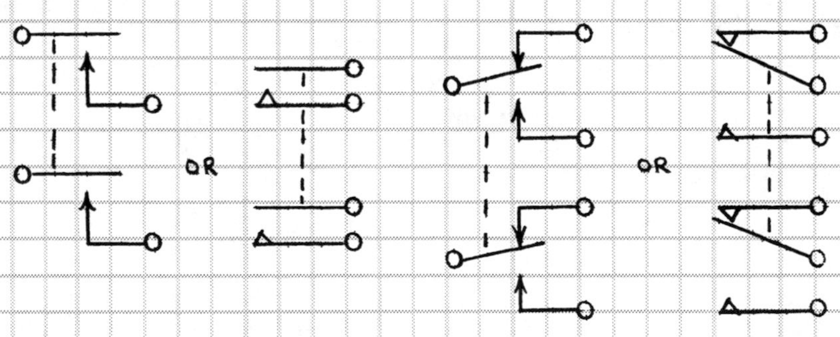

# MOTORS

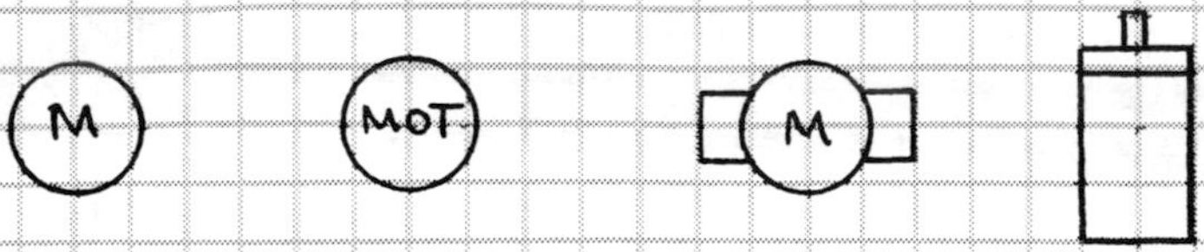

M

MOT

M

PHONO MOTOR

3-PHASE

4-PHASE

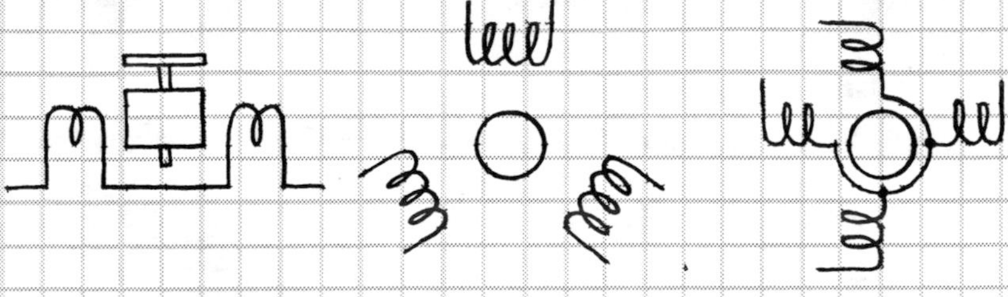

# SOLENOIDS

# METERS

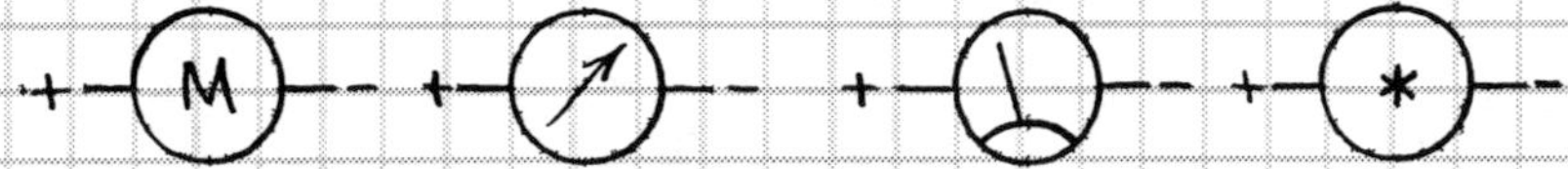

+ M −   +   −   +   −   + * −

*INSERT APPROPRIATE DESIGNATION (V=VOLTMETER; A = AMMETER; mA = MILLIAMMETER; ETC.)

# DELAY LINE

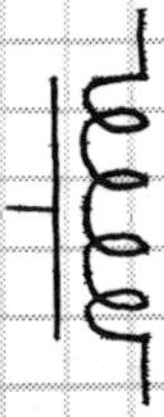

IN * OUT

* INSERT DELAY TIME.

# RESISTORS

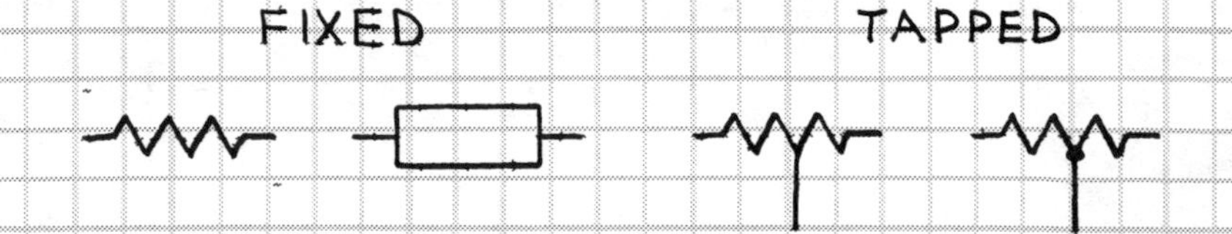


VARIABLE (POTENTIOMETERS, TRIMMERS, ETC.)

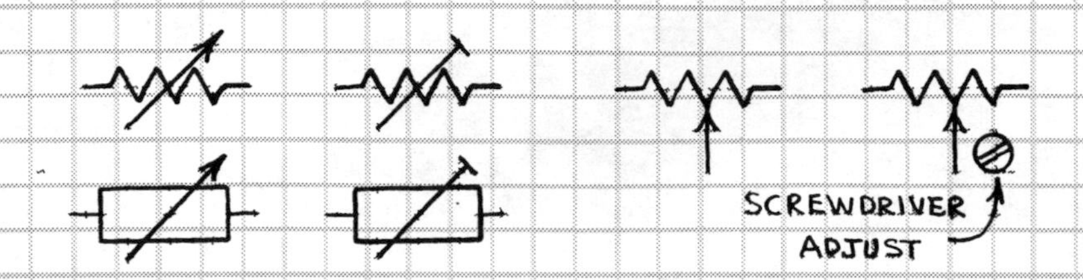


VOLTAGE DEPENDENT    CURRENT DEPENDENT

LIGHT DEPENDENT (PHOTORESISTORS)

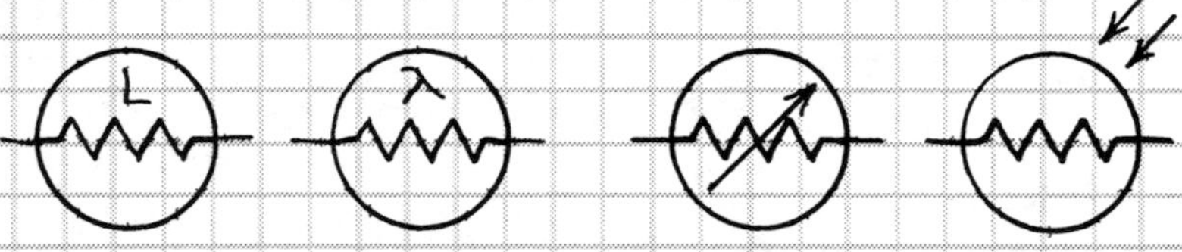


TEMPERATURE DEPENDENT (THERMISTORS)

# CAPACITORS

FIXED (NON-POLARIZED)

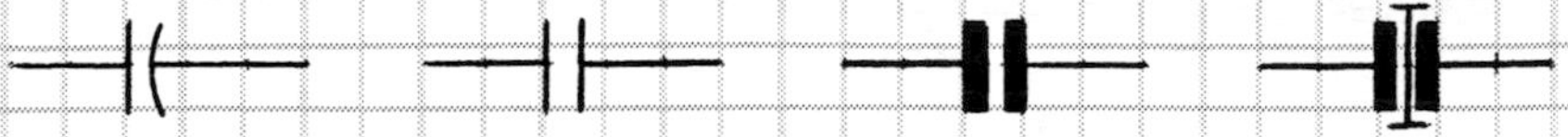

FIXED (POLARIZED)

+

+

+

+

VARIABLE

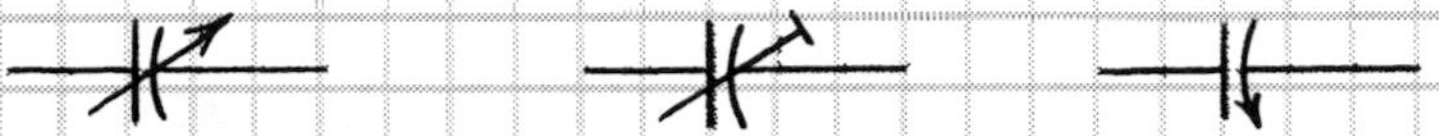

GANGED VARIABLE

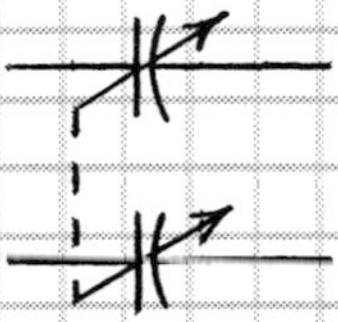

SPLIT STATOR

FEED THROUGH

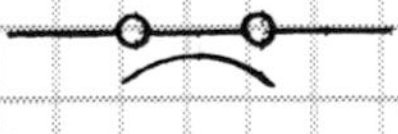

VOLTAGE VARIABLE (VARACTOR)

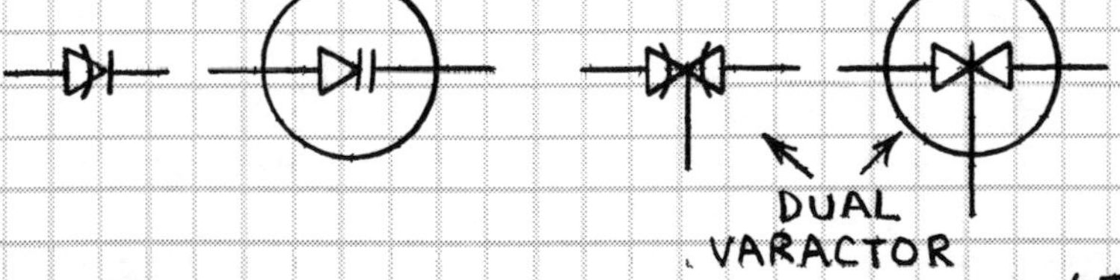

DUAL
VARACTOR

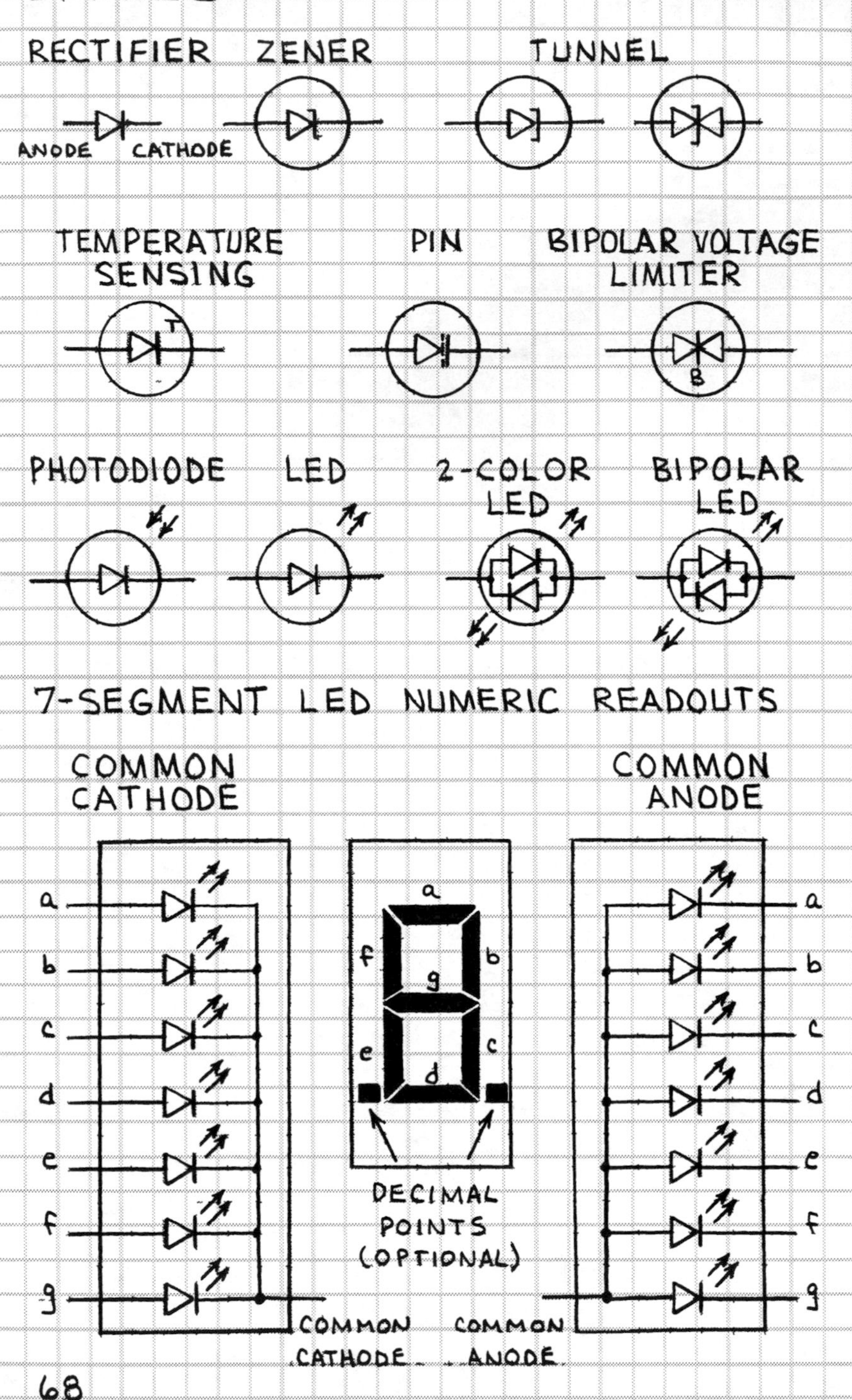

DIODES
RECTIFIER
ZENER
TUNNEL
ANODE
CATHODE
TEMPERATURE SENSING
PIN
BIPOLAR VOLTAGE LIMITER
T
B
PHOTODIODE
LED
2-COLOR LED
BIPOLAR LED
7-SEGMENT LED NUMERIC READOUTS
COMMON CATHODE
COMMON ANODE
a
b
c
d
e
f
g
DECIMAL POINTS (OPTIONAL)
COMMON CATHODE
COMMON ANODE

# 3-LAYER SWITCHES (DIACS)

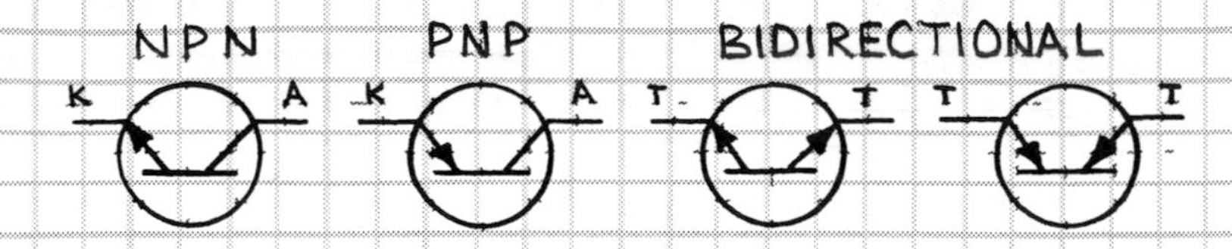


# 4-LAYER SWITCHES

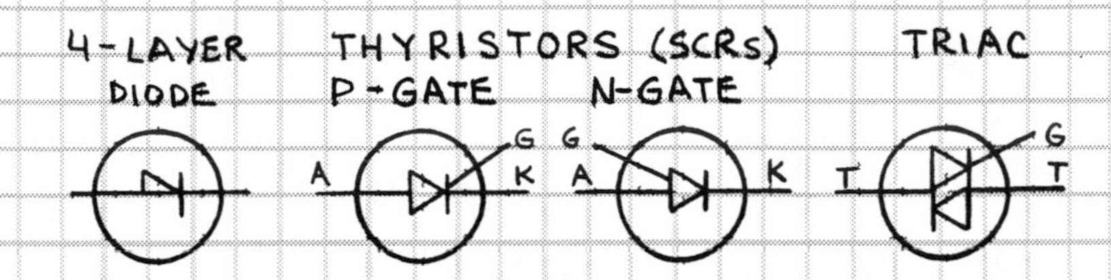


# TRANSISTORS

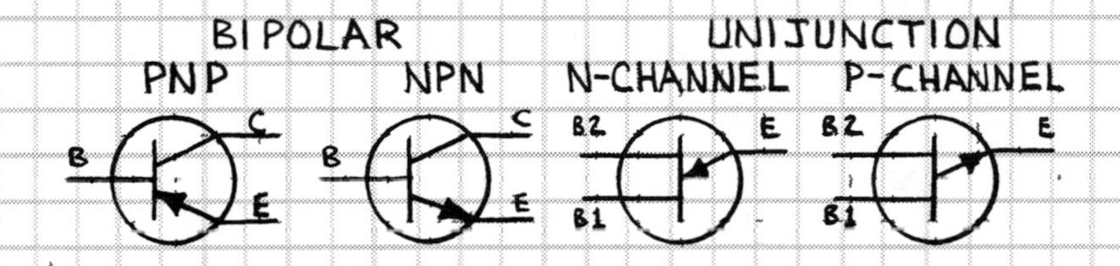


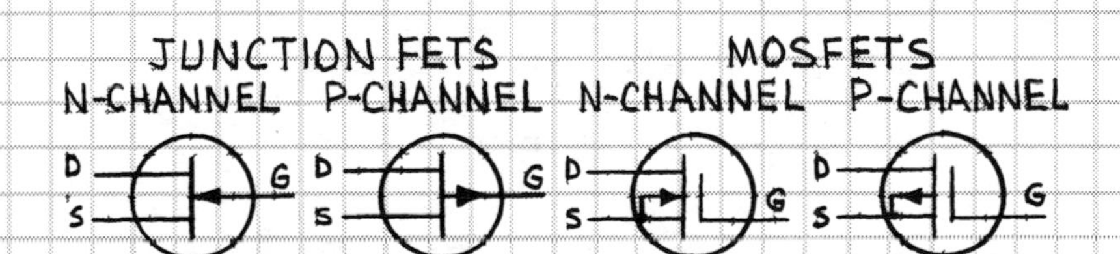


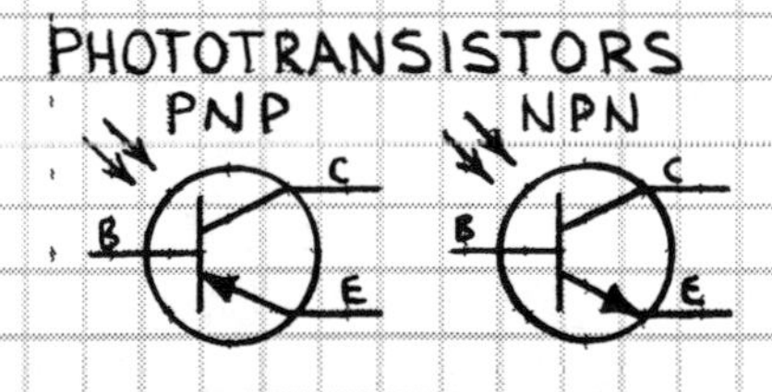


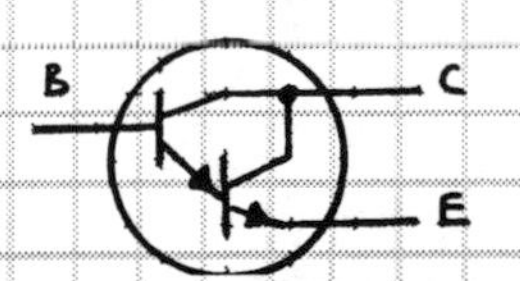

# ANALOG CIRCUITS

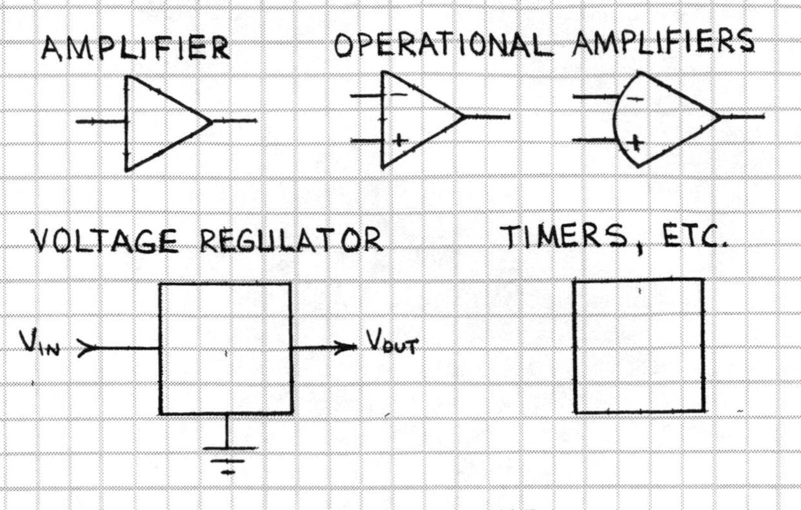


# CONVERTER CIRCUITS

DIGITAL-TO-ANALOG ANALOG-TO-DIGITAL

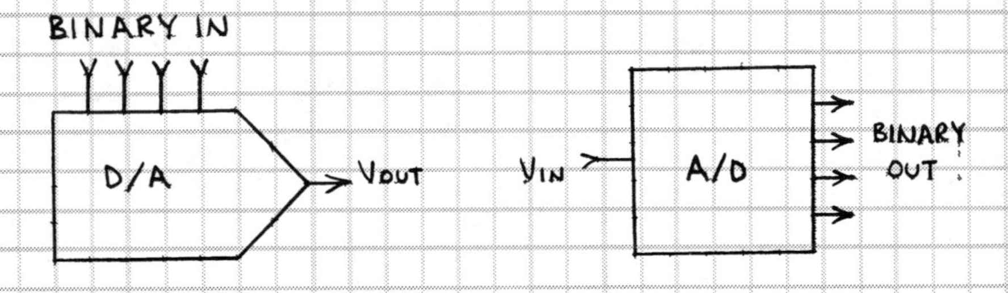


# DIGITAL DATA BUSES

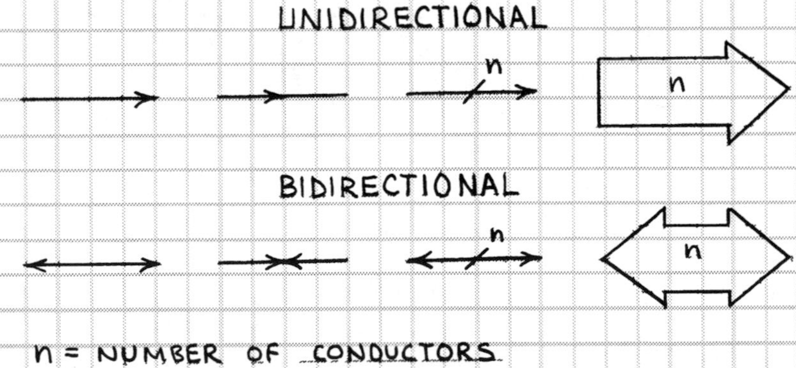


n = NUMBER OF CONDUCTORS

# DIGITAL CIRCUITS

## LOGIC GATES

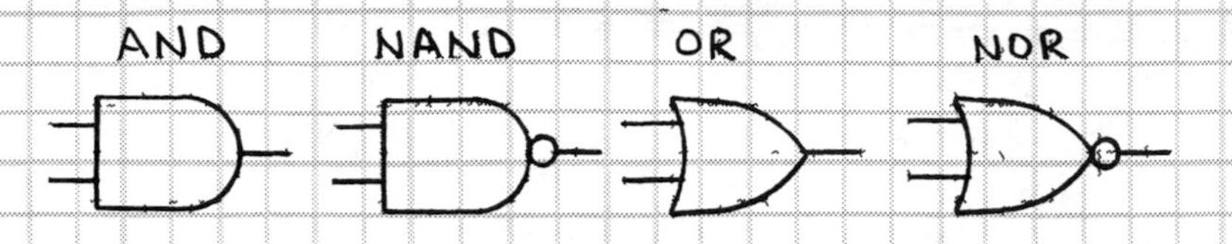


EXCLUSIVE OR EXCLUSIVE NOR INVERTERS

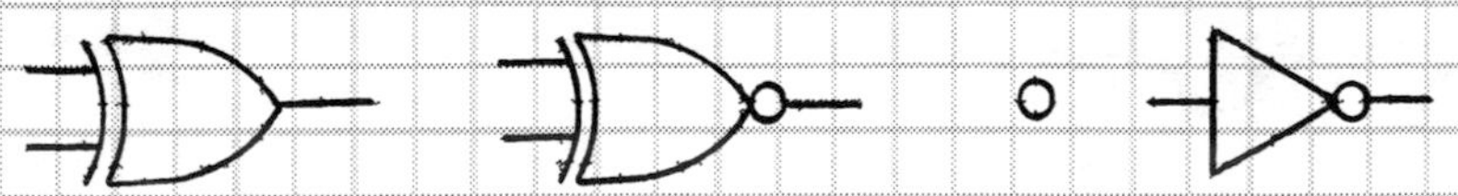

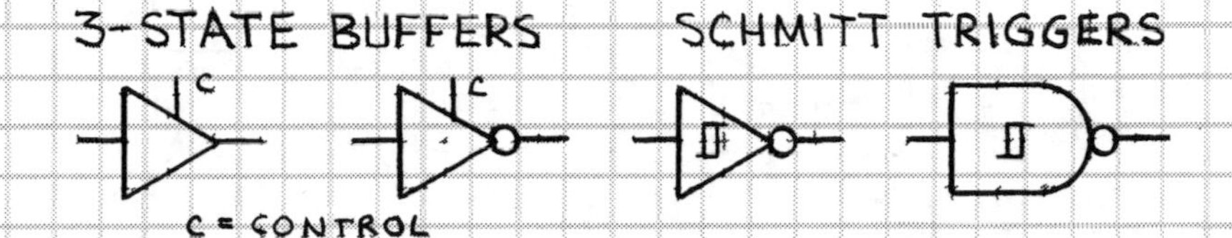


## FLIP-FLOPS

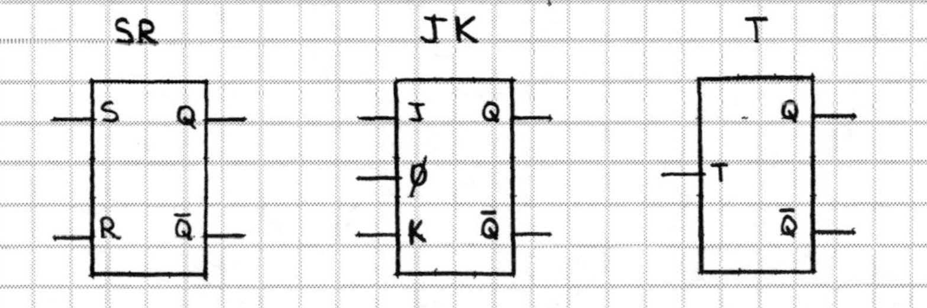


# COMPUTER FLOWCHART SYMBOLS

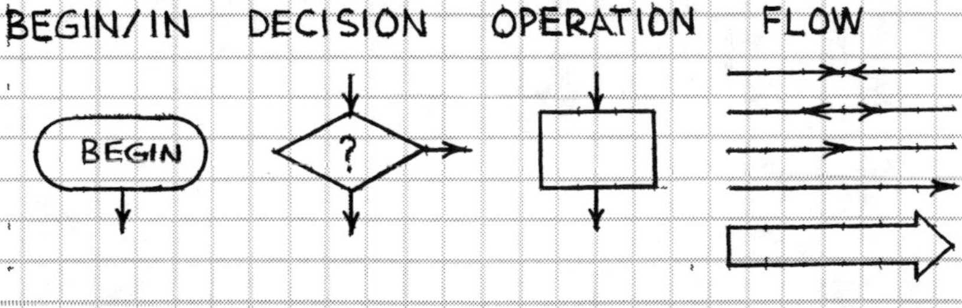

# 2. DEVICE PACKAGES

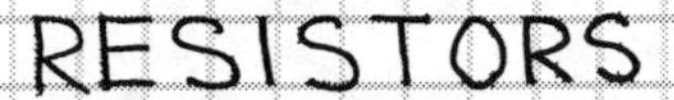

## RESISTORS

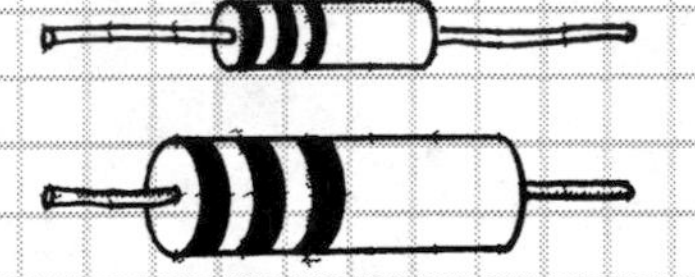

CARBON COMPOSITION          CARBON FILM

## CAPACITORS

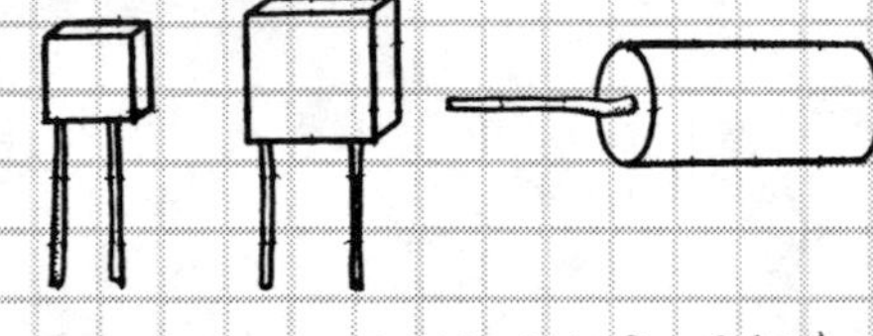

CERAMIC DISK          MOLDED   MULTILAYER CERAMIC

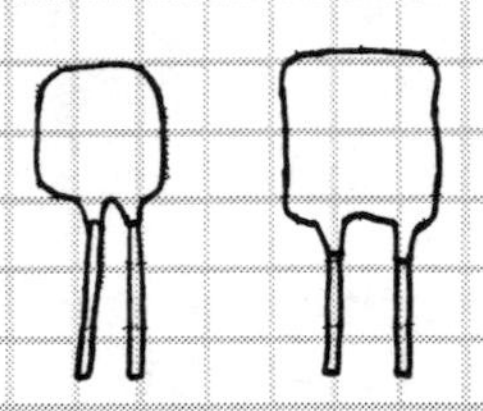

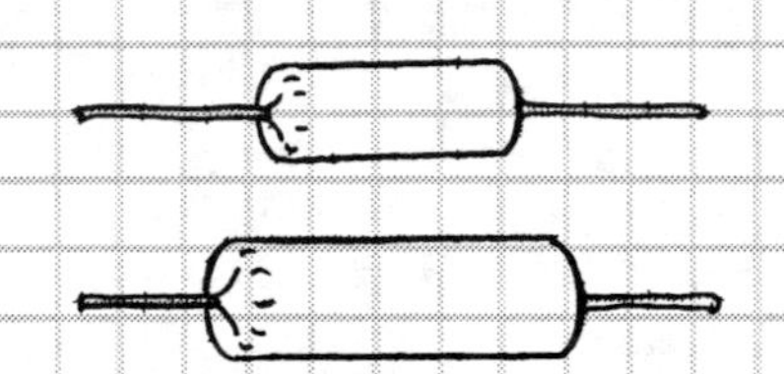

CONFORMALLY COATED MULTILAYER CERAMIC

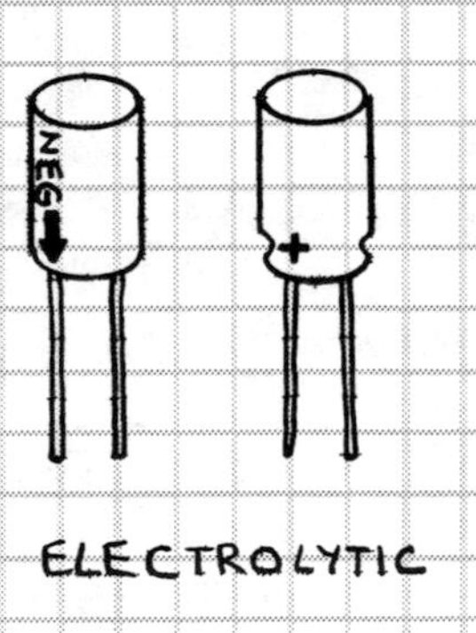


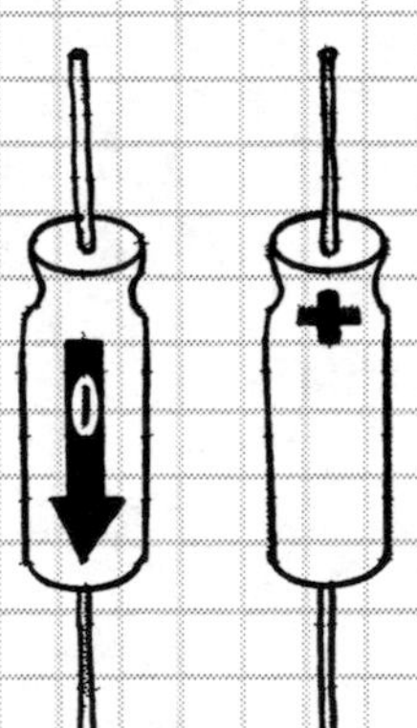

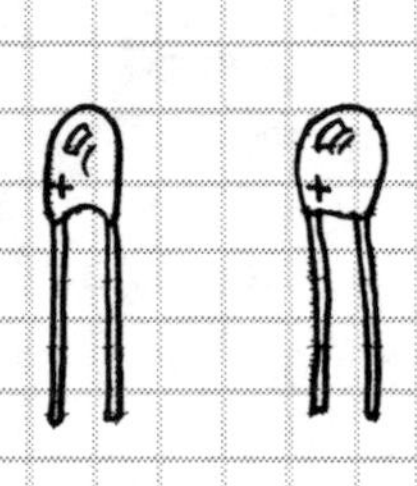

ELECTROLYTIC          DIPPED TANTALUM

# DIODES

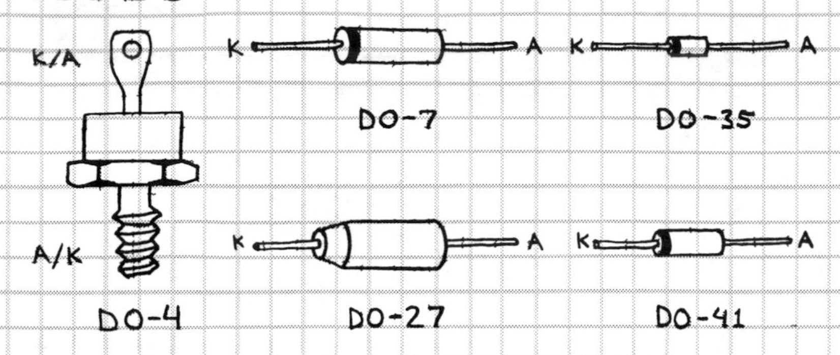


# BRIDGE RECTIFIERS

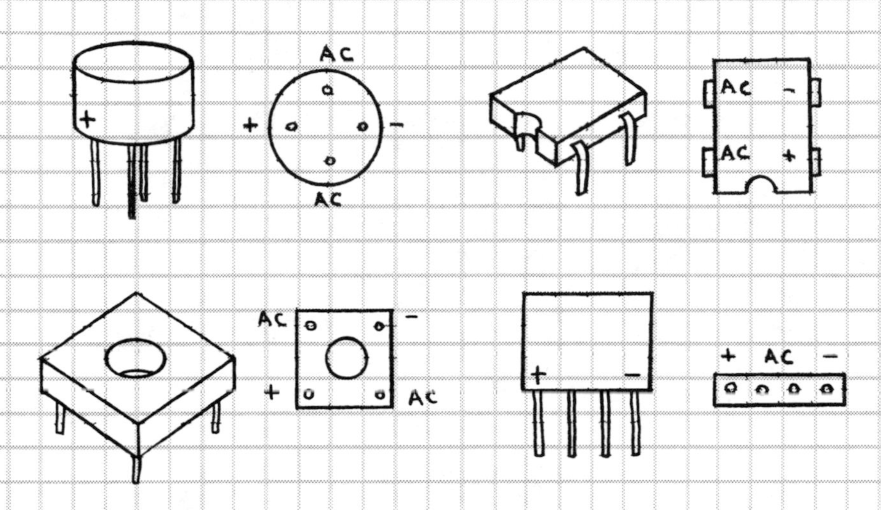


NOTE. ALWAYS CONSULT DEVICE SPECIFICATIONS TO VERIFY PIN IDENTIFICATION.

# LIGHT EMITTING DIODES

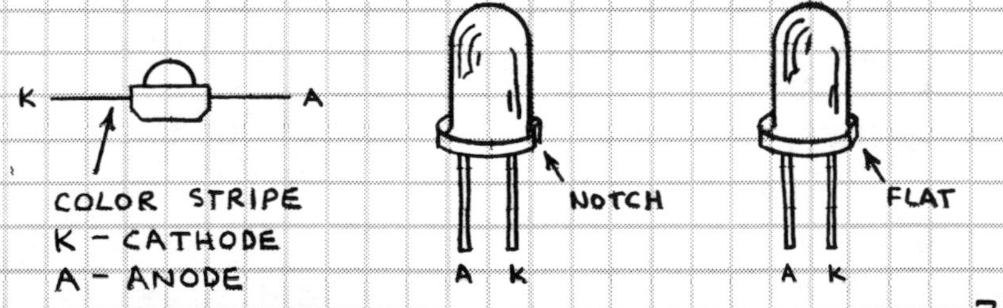

# TRANSISTORS (BOTTOM VIEW)

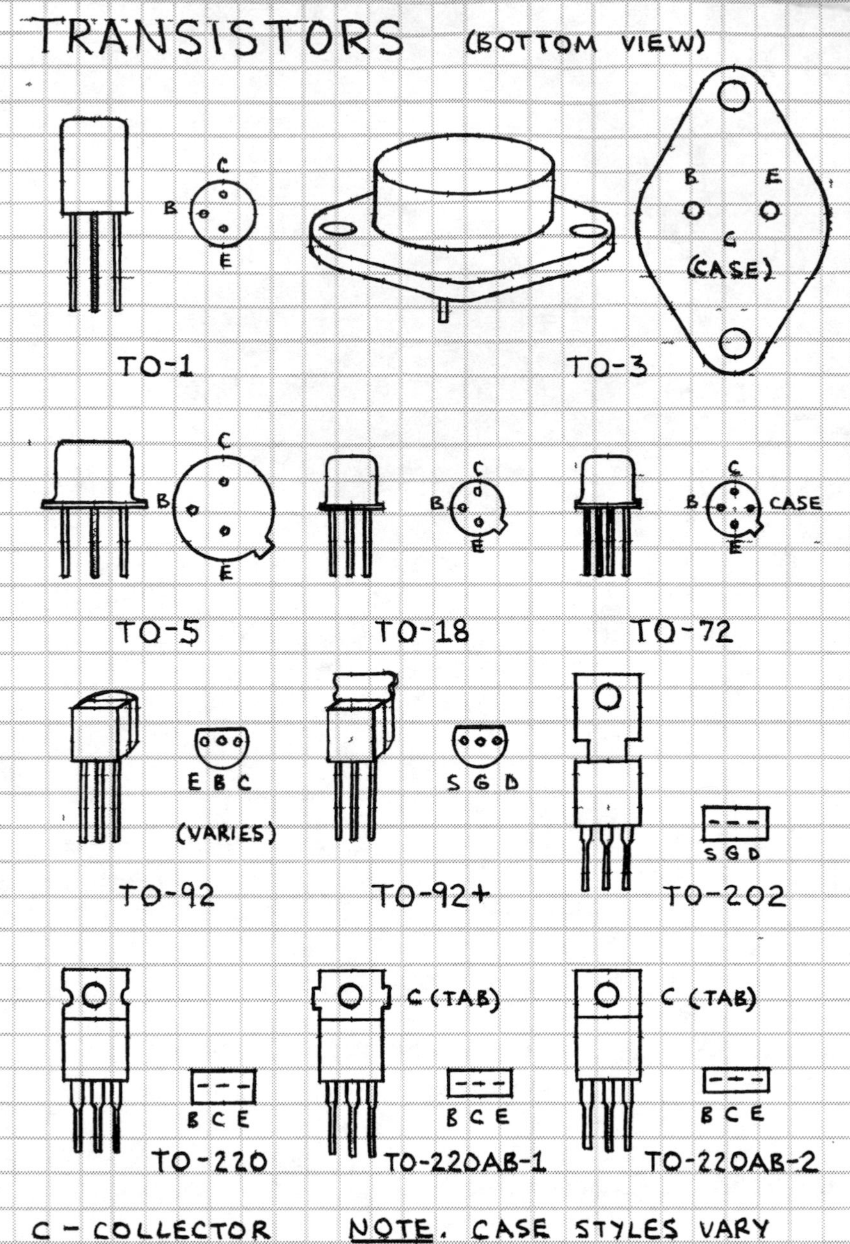


C - COLLECTOR
B - BASE
E - EMITTER
S - SOURCE
G - GATE
D - DRAIN

<u>NOTE</u>. CASE STYLES VARY AND MANY OTHERS ARE IN USE. ALWAYS CONSULT DEVICE SPECIFICATIONS TO VERIFY PIN IDENTIFICATION.

# OPTOCOUPLERS (TOP VIEW)

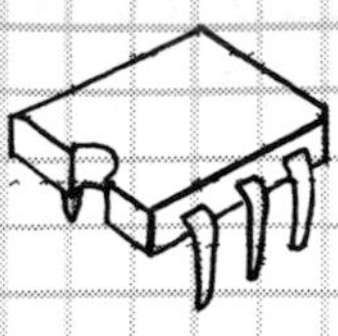

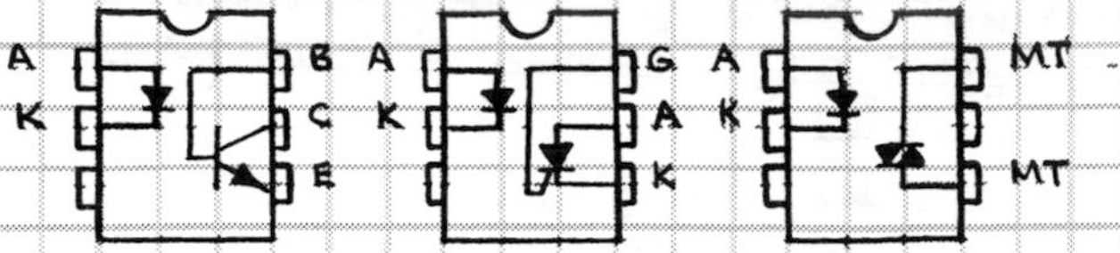

A K B C E — TRANSISTOR

A K G A K — SCR

A K MT MT — TRIAC

# INTEGRATED CIRCUITS (TOP VIEW)

8-PIN MINI-DIP

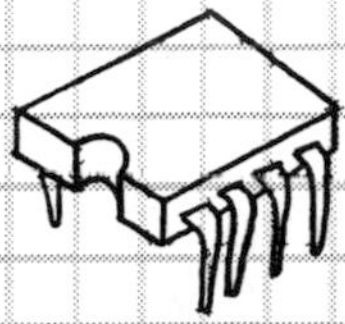

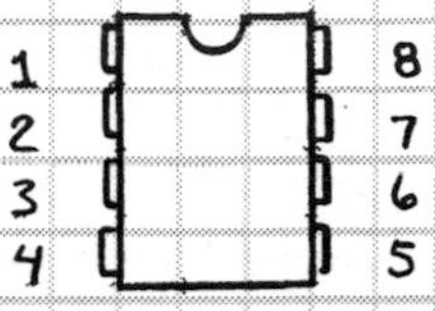

1 2 3 4 — 8 7 6 5

14-PIN DIP

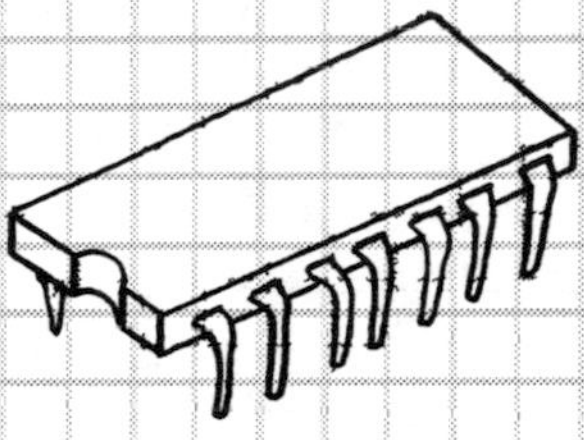

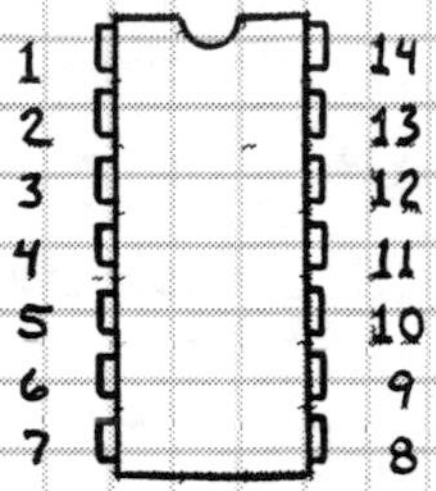

1 2 3 4 5 6 7 — 14 13 12 11 10 9 8

14-PIN SMALL OUTLINE (SO-14)

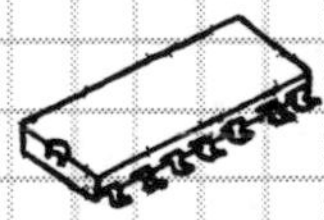

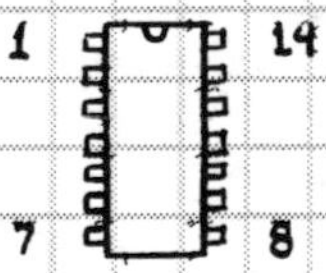

1 7 — 14 8

PLASTIC LEADED CHIP CARRIER (PLCC)

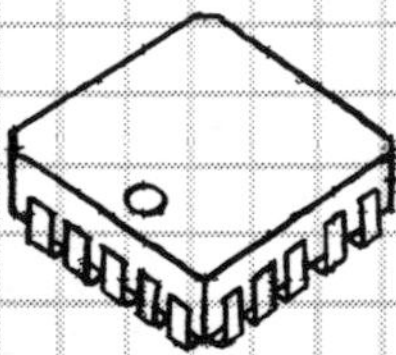

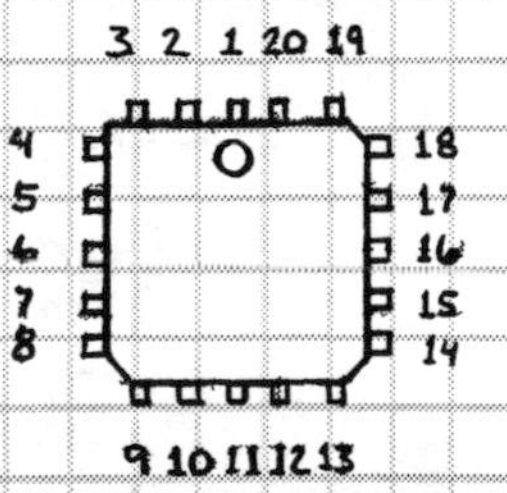

3 2 1 20 19

4 5 6 7 8 — 18 17 16 15 14

9 10 11 12 13

# BATTERIES

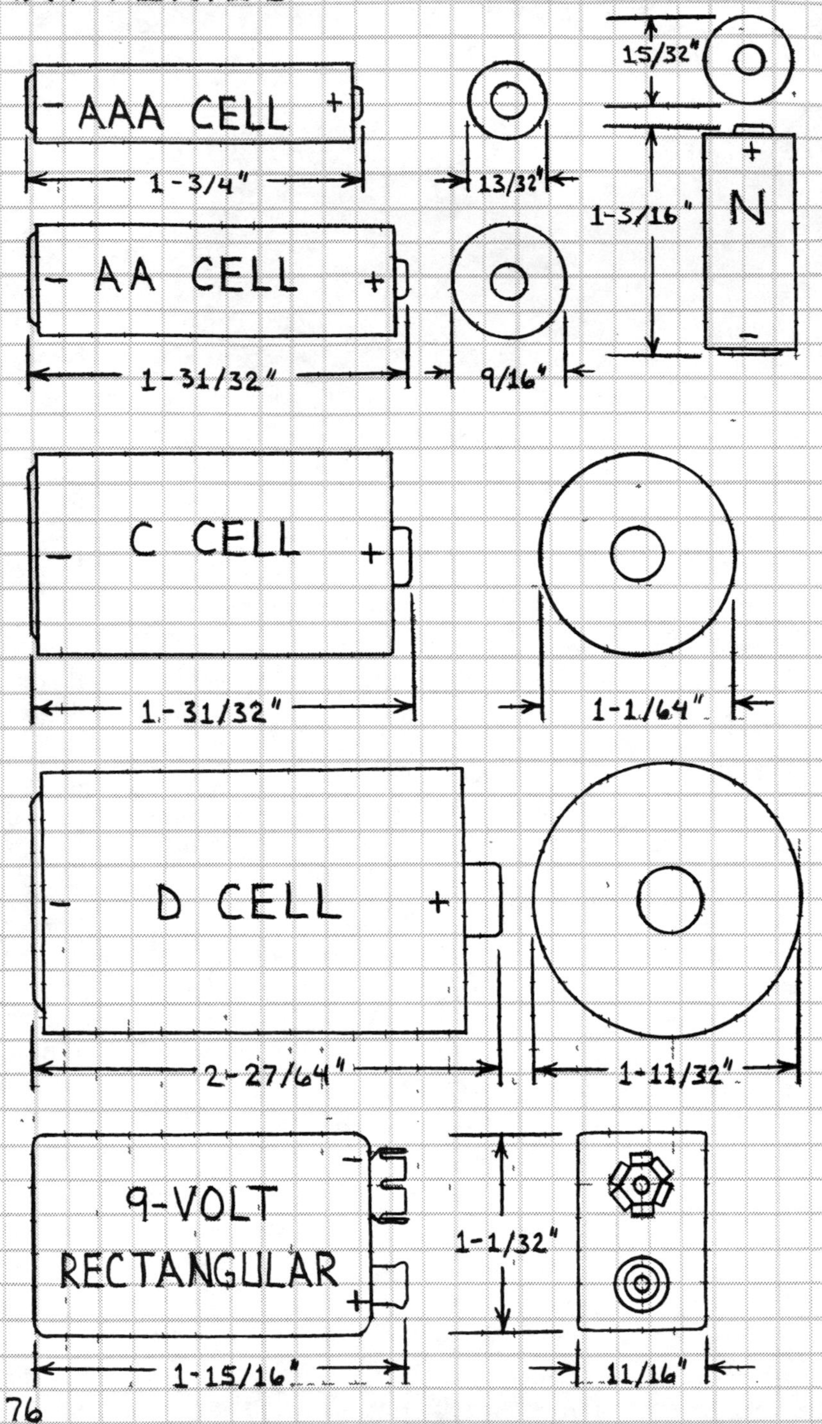

AAA CELL
1-3/4"
13/32"
15/32"
N
1-3/16"
AA CELL
1-31/32"
9/16"
C CELL
1-31/32"
1-1/64"
D CELL
2-27/64"
1-11/32"
9-VOLT
RECTANGULAR
1-15/16"
1-1/32"
11/16"

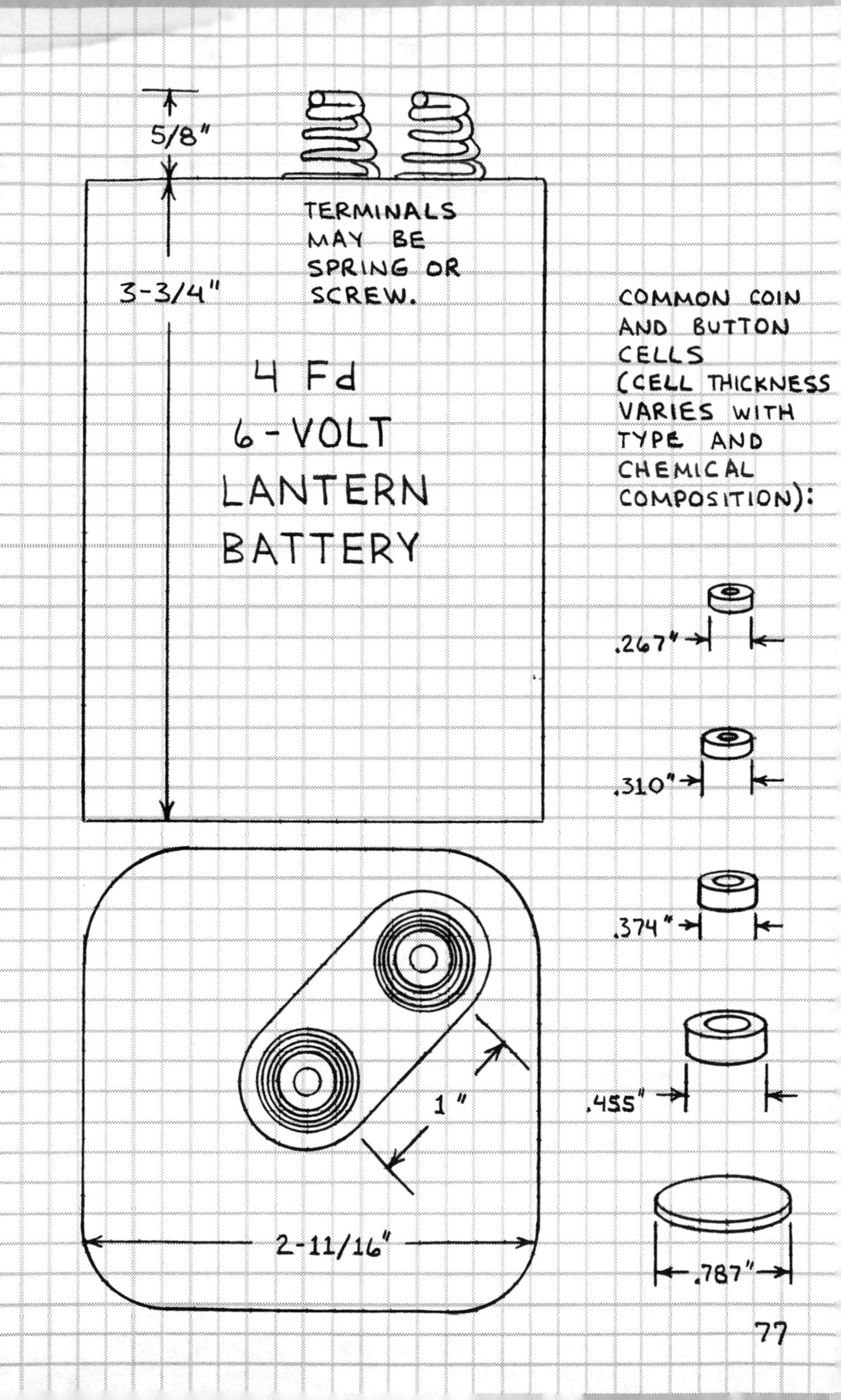
5/8"
TERMINALS MAY BE SPRING OR SCREW.
3-3/4"
4 Fd
6-VOLT
LANTERN
BATTERY
COMMON COIN AND BUTTON CELLS (CELL THICKNESS VARIES WITH TYPE AND CHEMICAL COMPOSITION):
.267"
.310"
.374"
.455"
1"
2-11/16"
.787"

# LAMPS

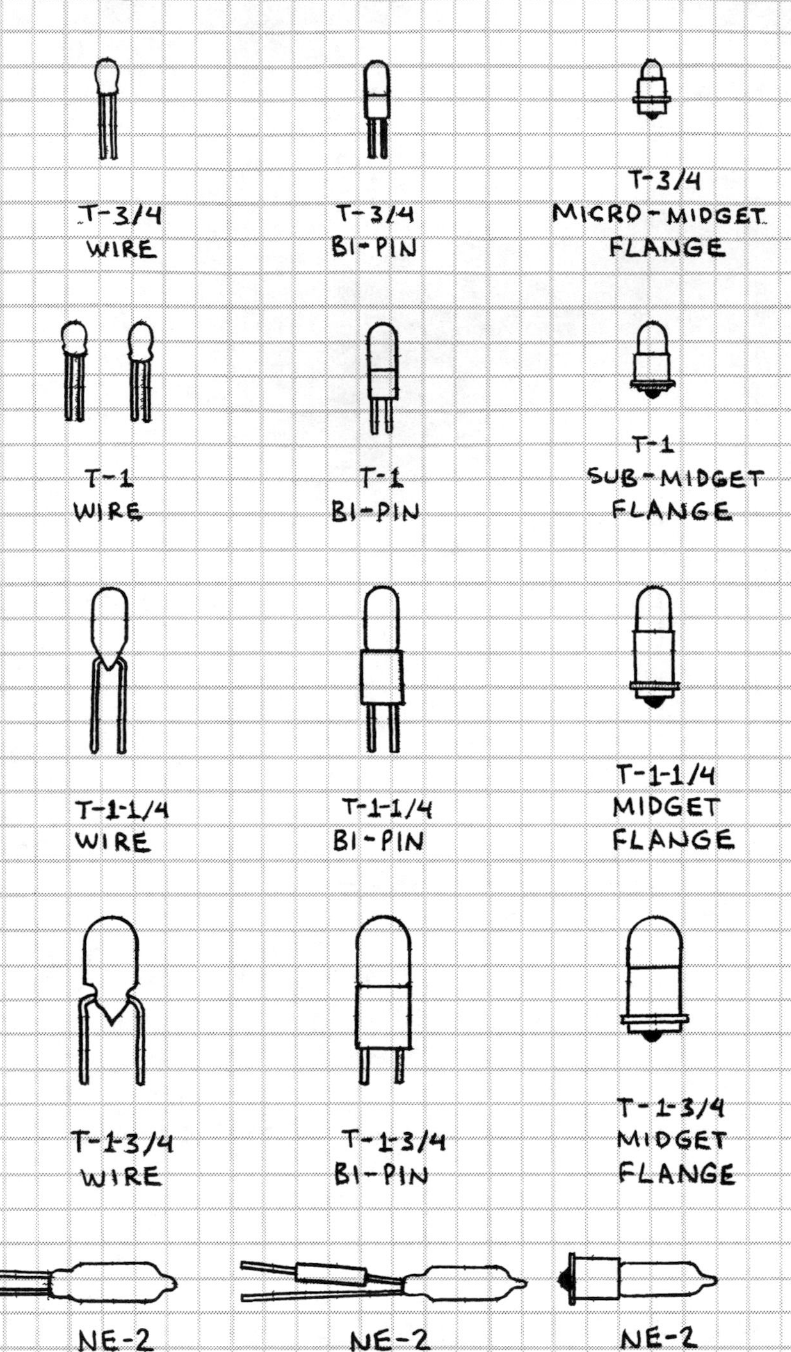
T-3/4
WIRE
T-3/4
BI-PIN
T-3/4
MICRO-MIDGET
FLANGE
T-1
WIRE
T-1
BI-PIN
T-1
SUB-MIDGET
FLANGE
T-1-1/4
WIRE
T-1-1/4
BI-PIN
T-1-1/4
MIDGET
FLANGE
T-1-3/4
WIRE
T-1-3/4
BI-PIN
T-1-3/4
MIDGET
FLANGE
NE-2
WIRE
NE-2
WIRE + RESISTOR
NE-2
FLANGE

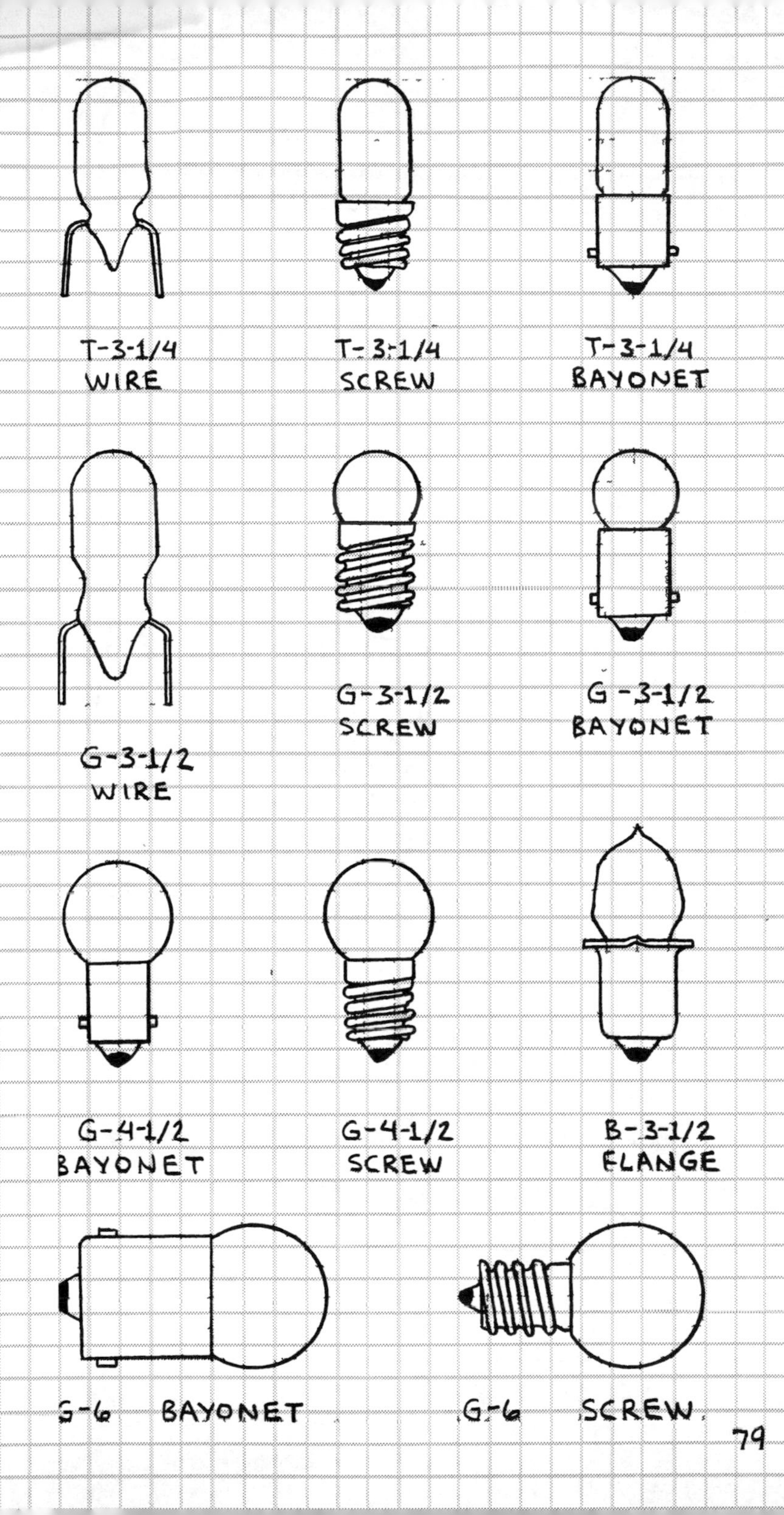
T-3-1/4
WIRE
T-3-1/4
SCREW
T-3-1/4
BAYONET
G-3-1/2
WIRE
G-3-1/2
SCREW
G-3-1/2
BAYONET
G-4-1/2
BAYONET
G-4-1/2
SCREW
B-3-1/2
FLANGE
G-6 BAYONET
G-6 SCREW

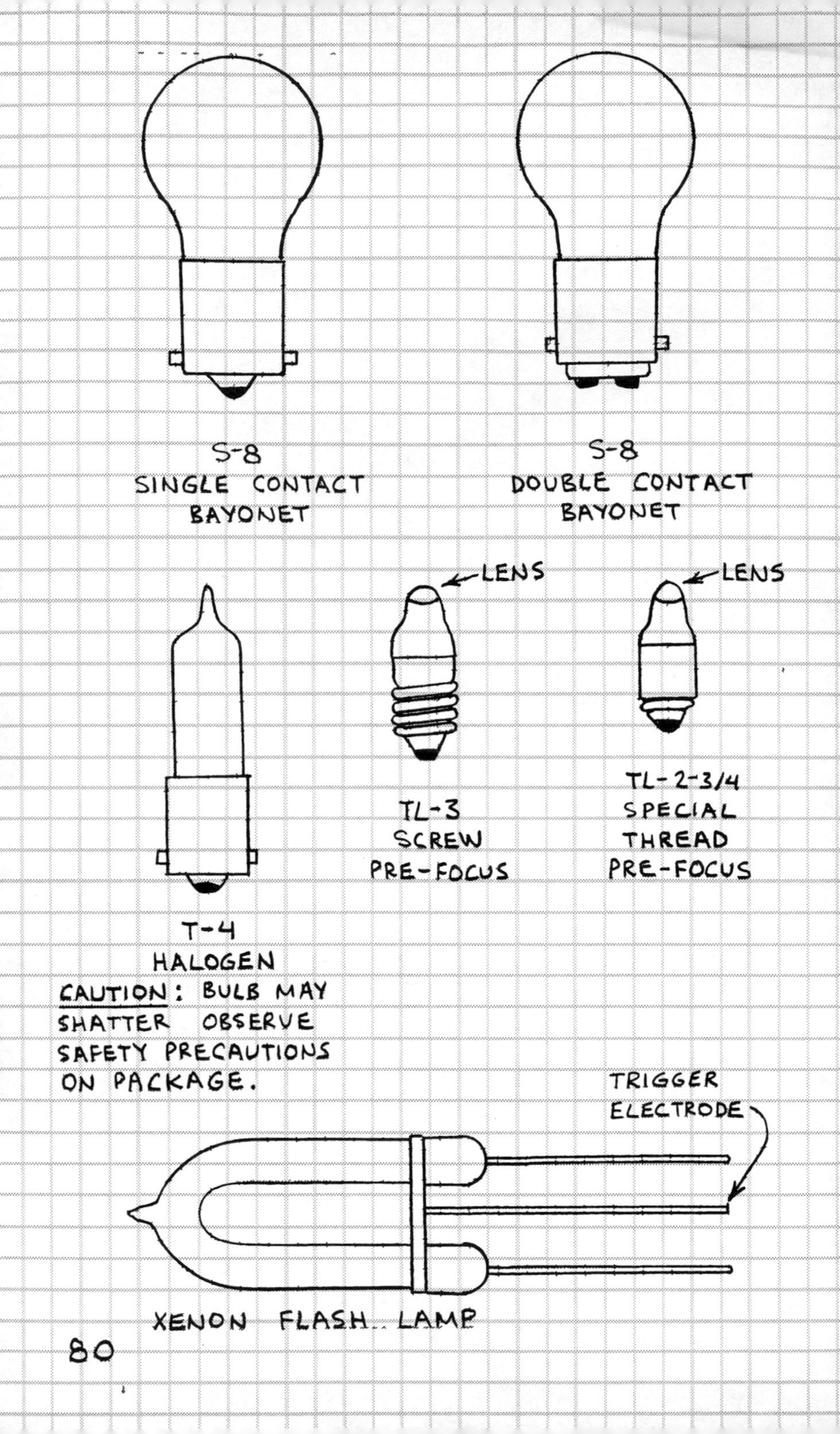
S-8
SINGLE CONTACT
BAYONET
S-8
DOUBLE CONTACT
BAYONET
LENS
LENS
TL-3
SCREW
PRE-FOCUS
TL-2-3/4
SPECIAL
THREAD
PRE-FOCUS
T-4
HALOGEN
CAUTION: BULB MAY
SHATTER OBSERVE
SAFETY PRECAUTIONS
ON PACKAGE.
TRIGGER
ELECTRODE
XENON FLASH LAMP

# 3. COMPONENT HANDLING

1. STORE COMPONENTS AT ROOM TEMPERATURE IN A DRY, DUST-FREE PLACE, PREFERABLY IN THE ORIGINAL PACKAGE.

2. AVOID DROPPING COMPONENTS. A FALL TO THE FLOOR SUBJECTS EVEN THE SMALLEST DEVICE TO MANY TIMES THE FORCE OF GRAVITY. A DROPPED DEVICE MAY APPEAR UNDAMAGED, BUT THE FORCE OF IMPACT MAY SEPARATE INTERNAL CONNECTIONS AND FORM TINY MICROCRACKS IN THE FUNCTIONAL PART OF THE DEVICE OR ITS PROTECTIVE COVERING OR COATING. CRACKS IN THE FUNCTIONAL PART OF THE DEVICE MAY RENDER IT USELESS, ALTER ITS SPECIFICATIONS OR DEGRADE ITS PERFORMANCE. CRACKS IN THE COATING WEAKEN THE DEVICE AND PERMIT THE ENTRY OF MOISTURE

3. AVOID OVERHEATING COMPONENTS WHEN SOLDERING OR DESOLDERING. PROTECT HEAT SENSITIVE COMPONENTS WITH A SOLDERING HEAT SINK OR PLIERS. COOL THESE COMPONENTS BY BLOWING ON THEM, BUT NOT THE CONNECTION, AFTER SOLDERING.

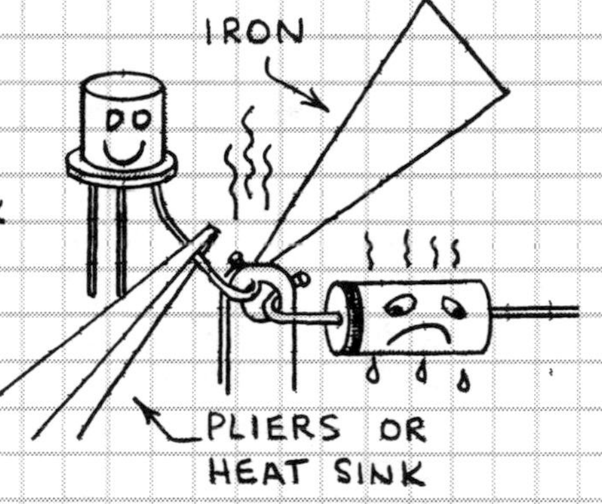


4. TO BEND A COMPONENT LEAD, GRASP THE LEAD WITH LONG NOSE PLIERS NEAR THE DEVICE AND THEN BEND THE LEAD WITH A FINGER. THE RADIUS OF THE BEND SHOULD EXCEED THE DIAMETER OF THE LEAD. BENDING LEADS WITHOUT PLIERS MAY FORM CRACKS BETWEEN LEAD AND DEVICE.

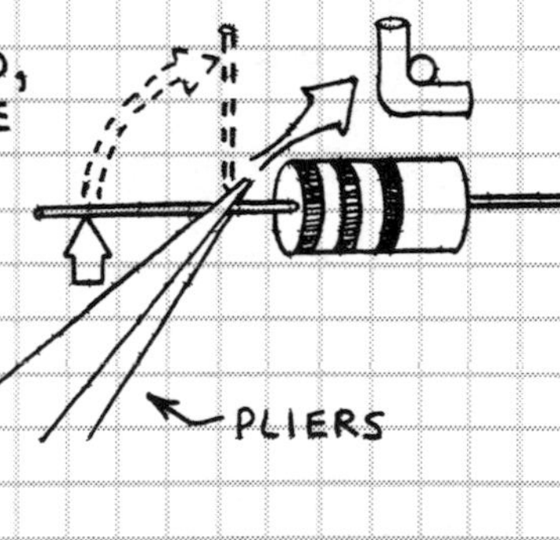

# ELECTROSTATIC DISCHARGE

IT IS WELL KNOWN THAT MOS (METAL-OXIDE-SEMICONDUCTOR) COMPONENTS CAN BE DAMAGED BY ELECTROSTATIC DISCHARGE (ESD). WHAT IS LESS WELL KNOWN IS THAT MANY OTHER COMPONENTS CAN ALSO BE DAMAGED BY ESD. COMPONENTS SUSCEPTABLE TO DAMAGE FROM ESD ARE SOMETIMES MARKED WITH A WARNING LABEL...

... BUT OFTEN THEY ARE NOT. THEREFORE IT IS IMPORTANT TO KNOW WHICH KINDS OF COMPONENTS ARE SUSCEPTABLE TO POSSIBLE DAMAGE FROM ESD.

ESD DAMAGE THRESHOLD OF CERTAIN COMPONENTS:

| EXTREMELY VULNERABLE (1 TO 1,000 V) | MODERATELY VULNERABLE (1,000 TO 5,000 V) | SOMEWHAT VULNERABLE (5,000 TO 15,000 V) |
|---|---|---|
| MOS TRANSISTORS<br>MOS ICs<br>μWAVE TRANSISTORS<br>JUNCTION FETs<br>LASER DIODES<br>METAL FILM RESISTORS | CMOS ICs<br>LS TTL ICs<br>SCHOTTKY TTL ICs<br>SCHOTTKY DIODES<br>LINEAR ICs | TTL ICs<br>SMALL SIGNAL DIODES AND TRANSISTORS<br>PIEZOELECTRIC CRYSTALS |
| THIS IS ONLY A PARTIAL LISTING. WHEN DOUBT EXISTS, TREAT SUSPECT DEVICES AS ESD SENSITIVE. | | |

TYPICAL ESD VOLTAGE GENERATED BY VARIOUS MATERIALS (75° F., 60% RELATIVE HUMIDITY):

| MATERIAL | ACTION | VOLTAGE |
|---|---|---|
| RUBBER COMB | STROKE DRY HAIR | -2,500 |
| DESK CHAIR | ROLL ACROSS PLASTIC FLOOR MAT | -2,000 |
| POLYETHYLENE BAG | CRUMPLE IN HAND | - 300 |
| TO-92 TRANSISTORS IN POLY BAG | SHAKE BAG SEVERAL TIMES | - 200 |
| PENCIL ERASER | RUB ACROSS CIRCUIT BOARD | + 100 |
| PLASTIC PARTS BOX | RUB WITH 100% COTTON FABRIC | + 100 |
| CLEAN PLASTIC TAPE (2" WIDE) | RAPIDLY UNROLL SEVERAL INCHES | + 500 |
| ADULT MALE (RUBBER SOLE SHOES) | WALK ACROSS CARPET | - 1,000 |

THESE MEASUREMENTS MADE WITH COMMERCIAL STATIC METER. ESD VOLTAGE IS FROM 10 TO 50 TIMES HIGHER WHEN RELATIVE HUMIDITY IS 10 TO 20%.

TYPICAL ESD DAMAGE TO GATE OF MOS FET:

# ESD HANDLING PRECAUTIONS

OBSERVE THE FOLLOWING PRECAUTIONS WHEN HANDLING COMPONENTS SUSCEPTABLE TO DAMAGE FROM ESD.

1. STORE COMPONENTS IN ORIGINAL PACKAGES, ELECTRICALLY CONDUCTIVE CONTAINERS OR CONDUCTIVE PLASTIC FOAM.

2. DO NOT TOUCH LEADS OR PINS.

3. DISCHARGE THE STATIC CHARGE ON YOUR BODY, BEFORE TOUCHING COMPONENTS, BY TOUCHING A GROUNDED METAL SURFACE (CABINET, APPLIANCE, ETC.).

4. PLACE COMPONENTS ON AN ALUMINUM FOIL SHEET OR TRAY OR ON CONDUCTIVE FOAM AFTER REMOVING THEM FROM THEIR CONTAINERS PRIOR TO INSTALLING THEM.

5. DO NOT SLIDE COMPONENTS ACROSS A WORK BENCH OR OTHER SURFACE.

6. KEEP STATIC-GENERATING MATERIALS (e.g.: PLASTIC, CELLOPHANE, CANDY WRAPPERS, PAPER, CARDBOARD, ETC.) AWAY FROM WORK AREA.

7. NEVER ALLOW CLOTHING TO MAKE CONTACT WITH COMPONENTS.

8. NEVER INSTALL ESD-SENSITIVE COMPONENTS IN A CIRCUIT WHEN POWER IS APPLIED, AND NEVER REMOVE COMPONENTS FROM A CIRCUIT WHEN POWER IS APPLIED.

9. WHEN POSSIBLE, USE A BATTERY-POWERED IRON TO MAKE SOLDER CONNECTIONS TO ESD-SENSITIVE COMPONENTS. AN AC-POWERED IRON MAY BE USED IF THE TIP DOES NOT CARRY STRAY VOLTAGE.

# 4. COMPONENT TESTING

ALTHOUGH COMPONENTS CONNECTED IN A CIRCUIT CAN BE TESTED, BETTER RESULTS ARE OBTAINED BY TESTING COMPONENTS NOT INSTALLED IN A CIRCUIT. SUGGESTED METHODS INCLUDE:

RESISTORS — MEASURE RESISTANCE WITH A MULTIMETER.

CAPACITORS — DISCHARGE CAPACITOR BY SHORTING LEADS. THEN CONNECT AN ANALOG MULTIMETER SET TO HIGHEST RESISTANCE RANGE ACROSS CAPACITOR. (BE SURE TO OBSERVE POLARITY OF ELECTROLYTIC CAPACITORS.) METER NEEDLE SHOULD MOVE TO RIGHT AND THEN FALL BACK TO INITIAL POINT. NEEDLE WILL MOVE MORE WITH LARGE VALUE CAPACITORS. IT MAY NOT MOVE WHEN VALUE IS BELOW 0.01 μF. IF NEEDLE REMAINS AT OR NEAR RIGHT SIDE OF METER, THE CAPACITOR IS SHORTED. IF NEEDLE FAILS TO MOVE, VALUE OF CAPACITOR IS BELOW 0.01 μF OR CAPACITOR IS OPEN.

DIODES — USE A MULTIMETER. RESISTANCE SHOULD BE LOW IN FORWARD DIRECTION AND HIGH IN REVERSE DIRECTION.

TRANSISTORS — THIS CIRCUIT PROVIDES A "GO/NO-GO" TEST FOR SWITCHING TRANSISTORS. RESPECTIVE LED GLOWS IF TRANSISTOR IS GOOD.

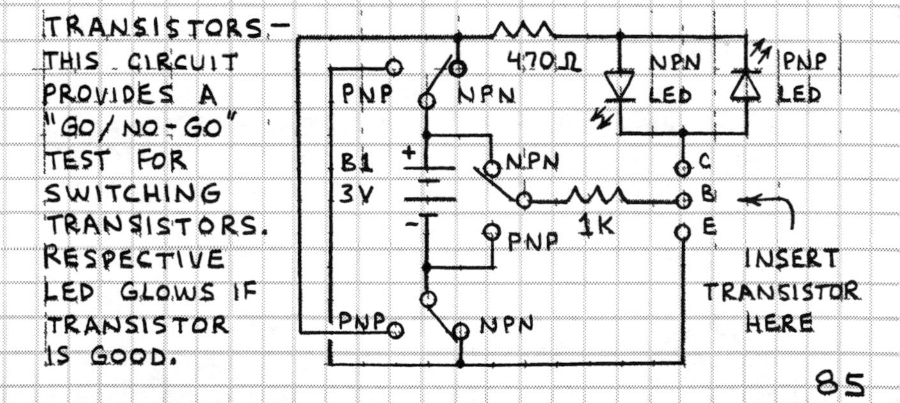

# 5. CIRCUIT DESIGN TIPS

1. USE EXISTING CIRCUITS AS BUILDING BLOCKS TO FORM ENTIRELY NEW CIRCUITS.

2. ALWAYS REVIEW THE MANUFACTURER'S SPECIFICATIONS FOR ACTIVE DEVICES (TRANSISTORS, INTEGRATED CIRCUITS, ETC.) BEFORE USING THEM IN A CIRCUIT. PAY PARTICULAR ATTENTION TO OPERATING VOLTAGES, INPUT AND OUTPUT REQUIREMENTS AND POTENTIAL PROBLEMS (SUCH AS OSCILLATION, NOISE, LATCHUP, ETC.).

3 BYPASS CAPACITORS, WHILE NOT ALWAYS REQUIRED, CAN PREVENT NOISE AND OSCILLATION IN ANALOG CIRCUITS AND FALSE TRIGGERING AND MEMORY LOSS IN DIGITAL CIRCUITS. IN ANALOG CIRCUITS PLACE A 0.1 μF AND 1.0 μF CAPACITOR ACROSS BATTERY LEADS WHERE THEY ENTER THE CIRCUIT BOARD. USE 0.1 μF CAPACITORS FROM POWER SUPPLY PINS OF OPERATIONAL AMPLIFIERS TO GROUND. IN DIGITAL CIRCUITS PLACE A 0.1 μF CAPACITOR ACROSS THE POWER SUPPLY PINS OF EACH CHIP.

4. COMPONENT SUBSTITUTION IS GENERALLY OKAY. HERE ARE SOME GENERAL GUIDELINES:

a. RESISTORS—USE NEXT CLOSEST VALUE. USE EQUAL OR HIGHER POWER RATING. CIRCUIT PERFORMANCE MAY BE ALTERED. FOR EXAMPLE, A SMALLER THAN SPECIFIED RESISTOR IN SERIES WITH AN LED WILL INCREASE CURRENT THROUGH THE LED

b. CAPACITORS—USE NEXT CLOSEST VALUE. USE EQUAL OR HIGHER VOLTAGE RATING. CIRCUIT PERFORMANCE MAY BE ALTERED FOR EXAMPLE, USING A SMALLER THAN SPECIFIED CAPACITOR IN A TIMER CIRCUIT WILL REDUCE THE TIMING CYCLE.

c. BIPOLAR TRANSISTORS—SUBSTITUTE WITHIN SAME FAMILY. OBSERVE POLARITY AND POWER.

# 6. CIRCUIT LAYOUT TIPS

1. CONNECTIONS BETWEEN COMPONENTS SHOULD BE AS SHORT AS POSSIBLE IN HIGH-SPEED DIGITAL CIRCUITS AND HIGH-FREQUENCY ANALOG CIRCUITS.

2. THE INPUT AND OUTPUT SECTIONS OF HIGH-GAIN AMPLIFIERS SHOULD BE PHYSICALLY ISOLATED FROM ONE ANOTHER. OTHERWISE INDUCTANCE BETWEEN THE INPUT AND OUTPUT WIRING MAY CAUSE A PORTION OF THE OUTPUT SIGNAL TO BE FED BACK TO THE INPUT. THE RESULT WILL BE SEVERE OSCILLATION.

3. POWER TRANSISTORS, ICS AND SOME OTHER COMPONENTS THAT BECOME WARM DURING OPERATION OFTEN PERFORM BETTER WITH A HEAT SINK. THEREFORE, LEAVE SPACE AROUND SUCH COMPONENTS FOR A HEAT SINK. AVOID PLACING HEAT SENSITIVE COMPONENTS NEAR COMPONENTS THAT MAY BECOME HOT.

4. USE INSULATED WIRE FOR INTERCONNECTIONS. INSULATE EXPOSED COMPONENT LEADS MOUNTED CLOSE TO OTHER EXPOSED LEADS OR HARDWARE.

5. ALL LEADS THAT CARRY HOUSEHOLD LINE CURRENT MUST BE INSULATED.

6. CIRCUITS IN WHICH A CURRENT FLOW IS SUDDENLY SWITCHED OFF OR ON MAY EMIT RADIO FREQUENCY RADIATION THAT CAN CAUSE SIGNIFICANT INTERFERENCE IN NEARBY RADIOS AND TELEVISIONS. RADIO FREQUENCY EMISSION CAN BE REDUCED BY ENCLOSING THE ENTIRE CIRCUIT IN A GROUNDED METAL ENCLOSURE. EXTERNAL CONNECTIONS TO OR FROM THE ENCLOSURE SHOULD BE MADE WITH SHIELDED CABLES.

7. USE STRANDED WIRE FOR ALL CONNECTIONS THAT ARE NOT FIXED IN POSITION (BATTERY CLIP LEADS, ETC.). USE SOLID WIRE FOR FIXED CONNECTIONS.

# 7. HEATSINKING

HEAT IS PRODUCED WHEN AN ELECTRICAL CURRENT FLOWS THROUGH A COMPONENT OR A CONDUCTOR. MOST COMPONENTS ARE SPECIFIED FOR OPERATION WITHIN A GIVEN TEMPERATURE RANGE. A HEATSINK WILL HELP REMOVE EXCESS HEAT FROM A COMPONENT. THERE ARE THREE PRIMARY MEANS BY WHICH HEAT LEAVES A COMPONENT:

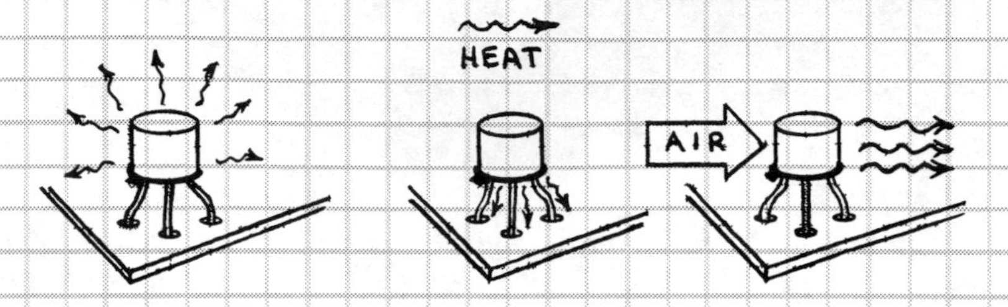


**RADIATION**

HEAT IS RADIATED INTO SPACE AS ELECTROMAGNETIC RADIATION.

**CONDUCTION**

HEAT IS CONDUCTED AWAY THROUGH DEVICE LEADS.

**CONVECTION**

HEAT IS CONDUCTED INTO SURROUNDING AIR AND WAFTED AWAY.

HEATSINKS ARE METAL STRUCTURES THAT IMPROVE THE EFFICIENCY WITH WHICH HEAT LEAVES A COMPONENT. THE THERMAL CONDUCTIVITY OF VARIOUS MATERIALS IS COMPARED BELOW.

| MATERIAL | CONDUCTIVITY (RELATIVE TO SILVER) |
|---|---|
| DIAMOND (II) | 5.4 |
| WATER | 1.4 |
| SILVER | 1.0 |
| COPPER | .93 |
| GOLD | .74 |
| ALUMINUM | .56 |
| NICKEL | .21 |
| IRON | .19 |
| TIN | .16 |
| MICA | .0014 |
| AIR | .000085 |

ALUMINUM IS THE MOST COMMON HEAT SINK MATERIAL. NOTE THAT COPPER IS NEARLY AS GOOD AS SILVER.

A HEATSINK WILL PERMIT A DEVICE SUCH AS A POWER SEMICONDUCTOR TO DISSIPATE AS MUCH AS TEN TIMES OR MORE HEAT THAN OTHERWISE. A HEATSINK WILL ALSO INCREASE A DEVICE'S RELIABILITY AND LIFETIME.

THE INTERFACE BETWEEN A HEATSINK AND A COMPONENT IS NOT PERFECTLY FLAT. THEREFORE A THERMALLY CONDUCTIVE PAD OR FILM OF SILICONE GREASE MUST BE PLACED BETWEEN THE HEATSINK AND THE DEVICE:

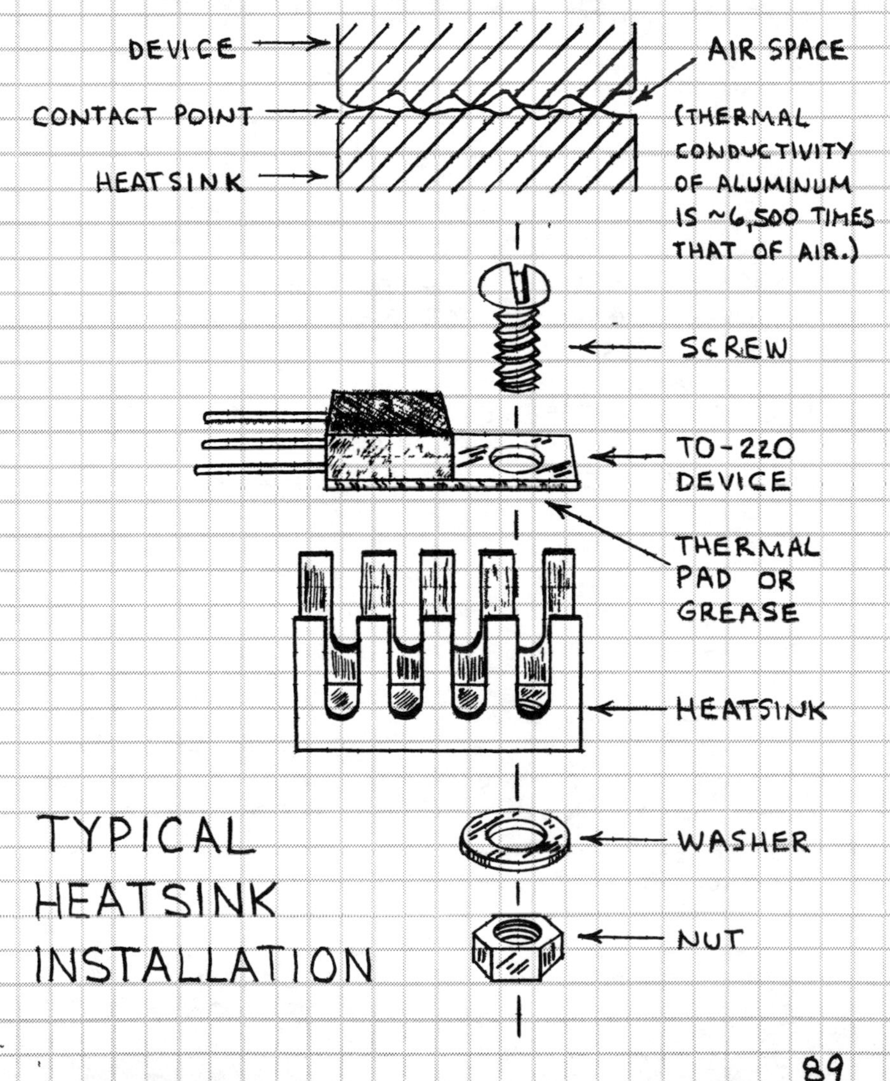


TYPICAL HEATSINK INSTALLATION

# 8. SOLDERING

FOLLOW THESE STEPS TO PRODUCE SUCCESSFUL SOLDER CONNECTIONS:

1. ELECTRONIC COMPONENTS AND CIRCUIT BOARDS CAN BE DAMAGED BY EXCESSIVE HEAT. THEREFORE, WHEN SOLDERING COMPONENTS TO A BOARD, ALWAYS USE A LOW-WATTAGE SOLDERING IRON (15 TO 40 WATTS). BE SURE TO TIN THE TIP ACCORDING TO THE INSTRUCTIONS SUPPLIED WITH THE IRON.

2. ALWAYS USE SMALL DIAMETER ROSIN CORE SOLDER WHEN SOLDERING ELECTRONIC PARTS. NEVER USE ACID CORE SOLDER. IT WILL CORRODE SOLDERED LEADS.

3. ALWAYS PREPARE THE SURFACES TO BE SOLDERED. SOLDER WILL NOT ADHERE TO PAINT, OIL, WAX, GREASE OR MELTED INSULATION. REMOVE THESE MATERIALS WITH A SOLVENT, STEEL WOOL OR FINE SANDPAPER. ALWAYS BUFF THE COPPER FOIL OF A CIRCUIT BOARD WITH STEEL WOOL. BE SURE THERE IS A GOOD CONNECTION BETWEEN SURFACES BEING SOLDERED.

4. TO SOLDER, HEAT THE CONNECTION FIRST, <u>NOT</u> THE SOLDER. AFTER A SECOND OR TWO TOUCH THE END OF A LENGTH OF SOLDER TO THE CONNECTION.

5. LEAVE THE HOT TIP OF THE IRON IN PLACE UNTIL MOLTEN SOLDER FLOWS THROUGH AND AROUND THE CONNECTION. THEN REMOVE THE IRON. IMPORTANT: DO NOT APPLY TOO MUCH SOLDER OR ALLOW THE CONNECTION TO MOVE BEFORE IT COOLS.

6. KEEP THE TIP OF THE IRON CLEAN AND SHINY. WIPE AWAY EXCESS SOLDER AND DEBRIS WITH A DAMP SPONGE OR CLOTH.

# DESOLDERING

A COMPONENT CAN BE REMOVED FROM A BOARD BY HEATING ITS CONNECTIONS WITH A HOT SOLDERING IRON UNTIL THE SOLDER MELTS AND THEN PULLING ON THE LEADS UNTIL THE COMPONENT IS FREE. UNLESS SPECIALIZED DESOLDERING TIPS ARE USED, THIS METHOD IS SUITABLE ONLY FOR INDIVIDUAL WIRES OR COMPONENTS WITH TWO LEADS. TO REMOVE COMPONENTS WITH MULTIPLE LEADS OR PINS, A DESOLDERING IRON OR TOOL SHOULD BE USED. FOLLOW THESE STEPS.

1. HEAT THE CONNECTION UNTIL THE SOLDER MELTS.

2. DESOLDERING IRON — SQUEEZE BULB BEFORE HEATING CONNECTION; RELEASE BULB WHEN SOLDER MELTS.

   DESOLDERING TOOL — SQUEEZE BULB OR ACTUATE PLUNGER. WHEN SOLDER MELTS, TOUCH TIP OF TOOL TO SOLDER AND RELEASE BULB OR PLUNGER. REPEAT IF NECESSARY.

   DESOLDERING BRAID — PLACE BRAID OVER SOLDER CONNECTION. PRESS BRAID AGAINST CONNECTION WITH TIP OF IRON UNTIL SOLDER MELTS AND FLOWS INTO BRAID.

3. REPAIR BROKEN AND SEPARATED FOIL PATTERN. SPLICES CAN BE MADE BY SOLDERING SHORT LENGTHS OF WIRE ACROSS BREAKS.

# SOLDERING PRECAUTIONS

1. A HOT SOLDERING IRON CAN CAUSE A FIRE OR BURN A FINGER. UNPLUG AN UNUSED SOLDERING IRON!

2. AVOID BREATHING SMOKE AND VAPOR FROM HOT SOLDER. SOLDER IN A WELL-VENTILATED AREA.

3. SUPERVISE CHILDREN WHO USE SOLDERING IRONS.

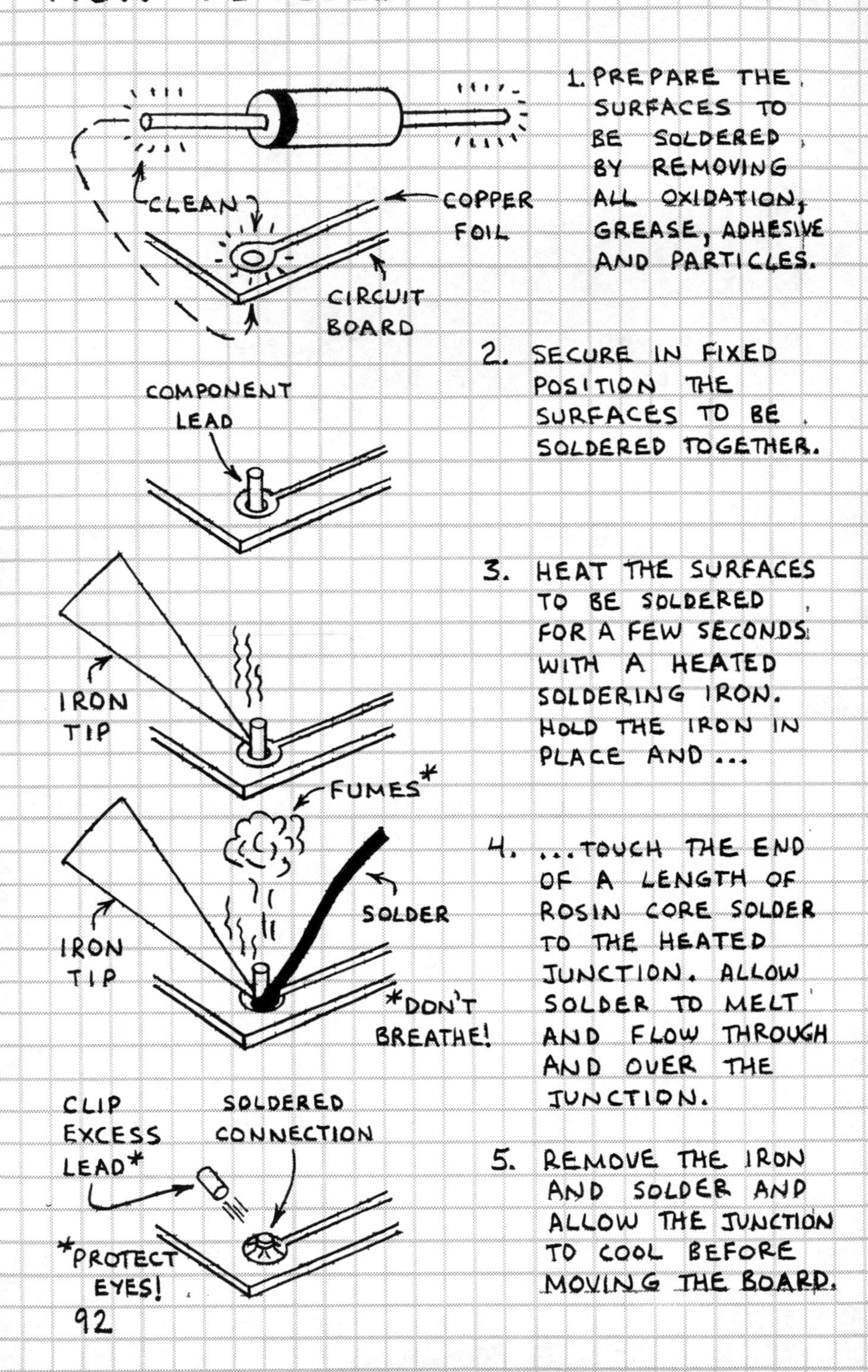
HOW TO SOLDER
CLEAN
COPPER FOIL
CIRCUIT BOARD
1. PREPARE THE SURFACES TO BE SOLDERED BY REMOVING ALL OXIDATION, GREASE, ADHESIVE AND PARTICLES.
2. SECURE IN FIXED POSITION THE SURFACES TO BE SOLDERED TOGETHER.
COMPONENT LEAD
IRON TIP
3. HEAT THE SURFACES TO BE SOLDERED FOR A FEW SECONDS WITH A HEATED SOLDERING IRON. HOLD THE IRON IN PLACE AND ...
FUMES*
IRON TIP
SOLDER
*DON'T BREATHE!
4. ...TOUCH THE END OF A LENGTH OF ROSIN CORE SOLDER TO THE HEATED JUNCTION. ALLOW SOLDER TO MELT AND FLOW THROUGH AND OVER THE JUNCTION.
CLIP EXCESS LEAD*
SOLDERED CONNECTION
*PROTECT EYES!
5. REMOVE THE IRON AND SOLDER AND ALLOW THE JUNCTION TO COOL BEFORE MOVING THE BOARD.

# HOW TO DESOLDER

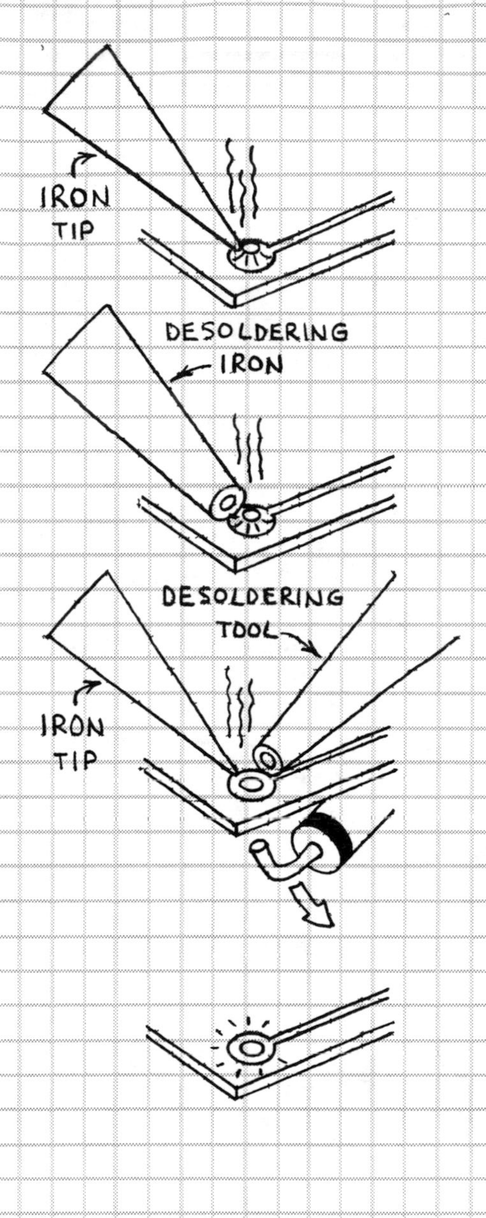


1. HEAT THE JUNCTION TO BE DESOLDERED WITH A HEATED SOLDERING IRON UNTIL THE SOLDER MELTS OR ...

2. ... HEAT THE JUNCTION WITH A HEATED DESOLDERING IRON UNTIL THE SOLDER MELTS.

3. SQUEEZE THE BULB OF A DESOLDERING TOOL (OR IRON), PLACE TIP OF TOOL (OR IRON) AS CLOSE AS POSSIBLE TO SOLDER AND RELEASE BULB. SOLDER WILL BE SLURPED UP INTO TOOL. COMPONENT LEAD CAN NOW BE REMOVED. NOTE THAT LEAD CAN BE REMOVED BY PULLING ON IT WHEN SOLDER IS MOLTEN.

4. CLEAN TERMINAL.

5. REPAIR BROKEN FOIL PATTERN WITH WIRE BRIDGE. SOLDER IN PLACE.

# 9. TROUBLESHOOTING

TROUBLESHOOTING IS THE PROCESS OF IDENTIFYING THE PROBLEM THAT CAUSES A CIRCUIT TO MALFUNCTION. WITH THE EXCEPTION OF MINOR PROBLEMS, TROUBLESHOOTING SOPHISTICATED SYSTEMS LIKE COMPUTERS AND VCRS IS BEST LEFT TO QUALIFIED TECHNICIANS. THE PROCEDURES LISTED BELOW CAN BE USED TO TROUBLESHOOT DO-IT-YOURSELF PROJECTS:

1. BE SURE YOU FULLY UNDERSTAND THE FUNCTION OF THE CIRCUIT AS DESCRIBED IN THE INSTRUCTIONS FOR ITS CONSTRUCTION.

2. IF THE CIRCUIT DOES NOT FUNCTION, BE SURE IT IS RECEIVING POWER. ARE THE BATTERIES FRESH AND INSTALLED CORRECTLY? ARE THE BATTERY HOLDER'S TERMINALS CLEAN? HAS A BATTERY CLIP LEAD BECOME BROKEN <u>INSIDE</u> ITS INSULATING JACKET? IS THE POWER CORD INSERTED IN AN OUTLET? IS A FUSE BLOWN? DOES THE CIRCUIT'S POWER REQUIREMENT EXCEED THE AVAILABLE POWER?

3. CAREFULLY COMPARE THE CIRCUIT WITH THE SCHEMATIC. HAS EVERY CONNECTION BEEN MADE? ARE ANY CONNECTIONS INCORRECT? ARE ANY SOLDER CONNECTIONS DEFECTIVE?

4. ARE POLARITY-SENSITIVE COMPONENTS LIKE ELECTROLYTIC CAPACITORS, DIODES AND TRANSISTORS INSTALLED CORRECTLY? ARE INTEGRATED CIRCUITS INSTALLED CORRECTLY?

5. ARE UNUSED INPUTS OF DIGITAL LOGIC CHIPS CONNECTED TO GROUND OR ONE SIDE OF THE POWER SUPPLY?

6. FOR BEST RESULTS FOLLOW AN ORGANIZED, LOGICAL APPROACH TO TROUBLESHOOTING. THE TROUBLESHOOTING TREE ON THE FACING PAGE ILLUSTRATES THIS APPROACH.

# TROUBLESHOOTING TREE

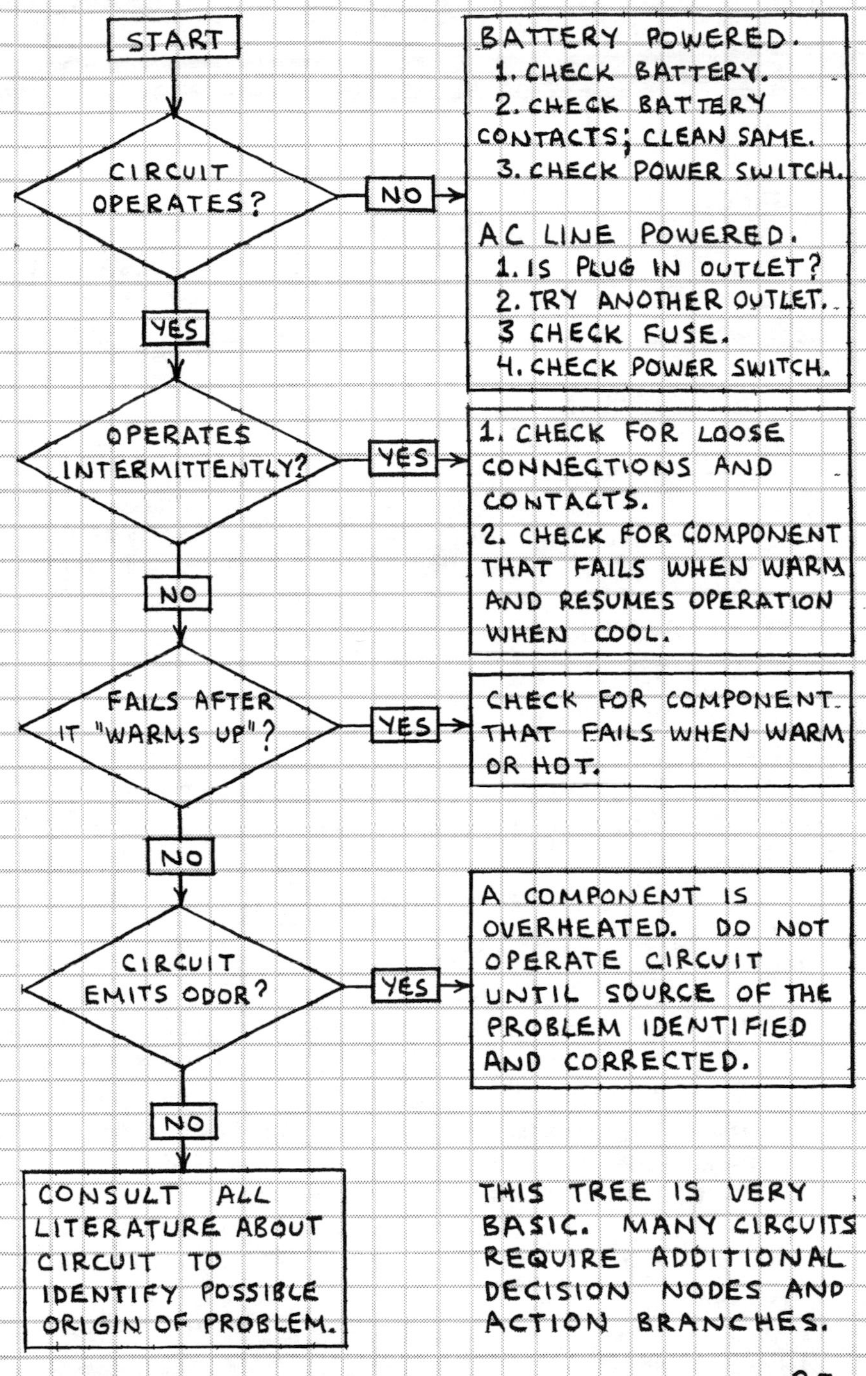

# DIGITAL TROUBLESHOOTING

THESE SIMPLE CIRCUITS PERMIT DIGITAL LOGIC CIRCUITS TO BE TESTED. <u>BOTH</u> CIRCUITS CAN BE ASSEMBLED USING SAME 4049.

## BOUNCELESS SWITCH

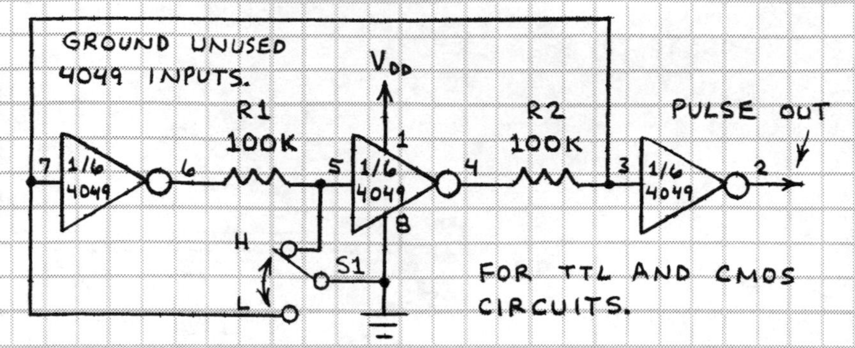


CONNECT $V_{DD}$ AND GROUND TO, RESPECTIVELY, POSITIVE SUPPLY AND GROUND OF THE CIRCUIT BEING TESTED. TOGGLE S1 TO PRODUCE CLEAN, NOISE-FREE PULSE.

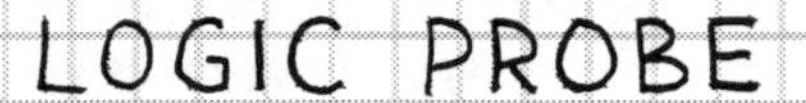

## LOGIC PROBE

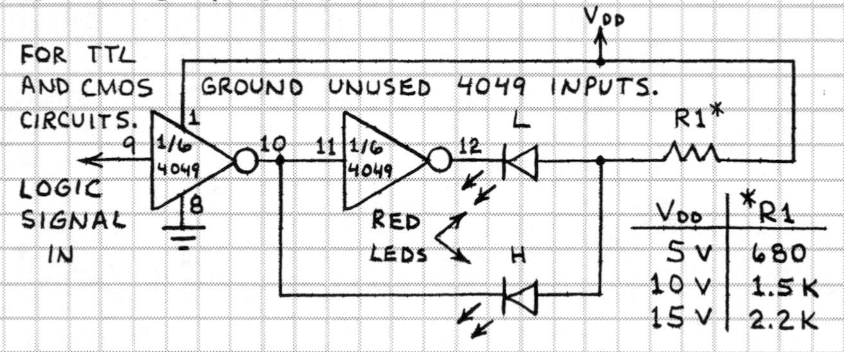


| $V_{DD}$ | *R1 |
|---|---|
| 5 V | 680 |
| 10 V | 1.5 K |
| 15 V | 2.2 K |

CONNECT $V_{DD}$ AND GROUND TO, RESPECTIVELY, POSITIVE SUPPLY AND GROUND OF THE CIRCUIT BEING TESTED. TOUCH INPUT PROBE TO TERMINAL OF CIRCUIT BEING TESTED. LEDs INDICATE LOGIC STATUS (L=LOW; H=HIGH). R1- TABLE GIVES VALUES FOR ~5 mA CURRENT. OKAY TO USE 2.2K FOR ALL VALUES OF $V_{DD}$ IF LEDs ARE SUPER-BRIGHT UNITS.

# ANALOG TROUBLESHOOTING

THESE CIRCUITS CAN BE USED TO TROUBLESHOOT AUDIO AMPLIFIERS AND TO DETERMINE THE CONTINUITY OF MULTI-CONDUCTOR WIRE AND CABLE. (SEE SAFETY PRECAUTIONS ON FOLLOWING PAGE.)

## SIGNAL INJECTOR

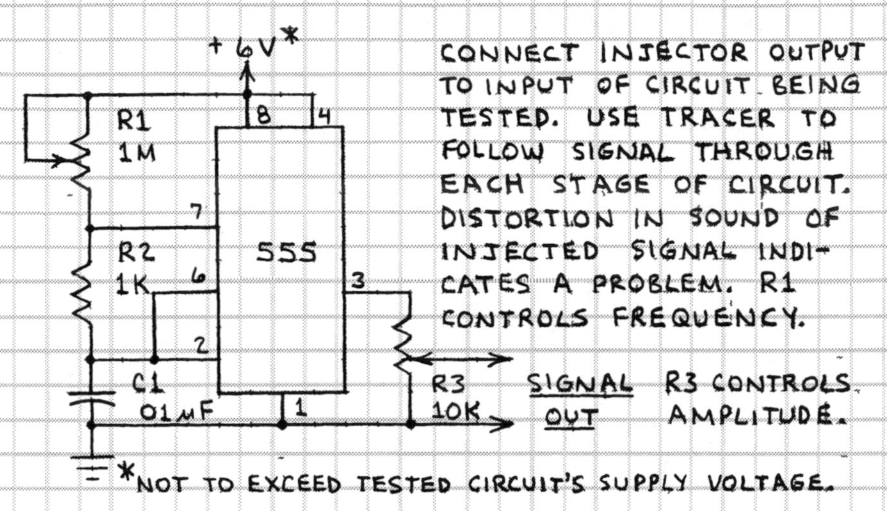


CONNECT INJECTOR OUTPUT TO INPUT OF CIRCUIT BEING TESTED. USE TRACER TO FOLLOW SIGNAL THROUGH EACH STAGE OF CIRCUIT. DISTORTION IN SOUND OF INJECTED SIGNAL INDICATES A PROBLEM. R1 CONTROLS FREQUENCY.

R3 CONTROLS AMPLITUDE.

*NOT TO EXCEED TESTED CIRCUIT'S SUPPLY VOLTAGE.

## SIGNAL TRACER

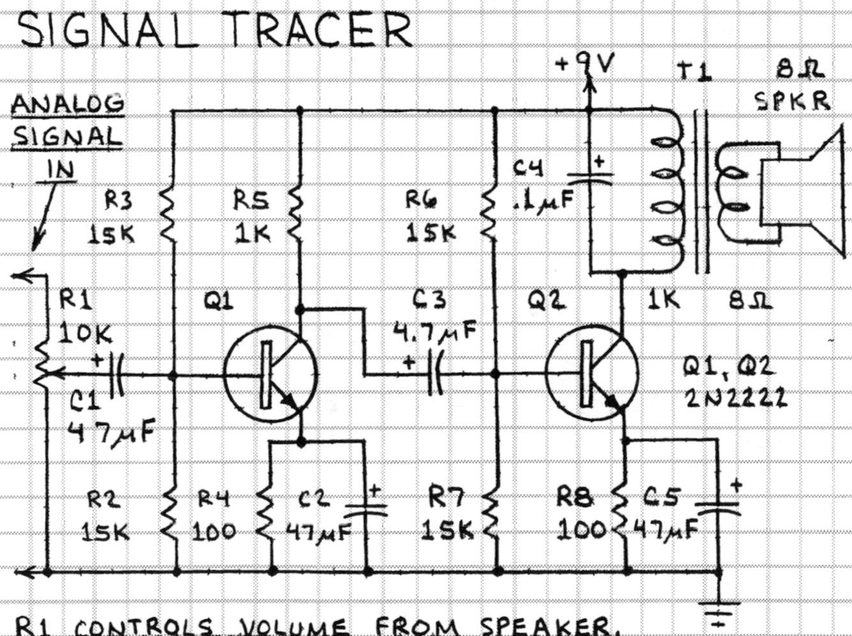


R1 CONTROLS VOLUME FROM SPEAKER.

# 10. SAFETY PRECAUTIONS

ELECTRONIC CIRCUITS POWERED BY HOUSEHOLD LINE CURRENT AND SOME BATTERY-POWERED CIRCUITS CAN CAUSE DANGEROUS ELECTRICAL SHOCKS. AN ELECTRICAL SHOCK CAN CAUSE HEART FAILURE. A SHOCK CAN ALSO CAUSE A VIOLENT MUSCLE REFLEX THAT MAY INJURE AN ARM OR LEG OR EVEN THROW YOU TO THE FLOOR. OBSERVE THESE PRECAUTIONS

1. HOUSEHOLD LINE CURRENT CAN KILL! ONLY EXPERIENCED TECHNICIANS SHOULD WORK ON A LINE-POWERED CIRCUIT WITH THE POWER ON!

2. EXPERIENCED TECHNICIANS NEVER WORK ALONE AND ALWAYS KEEP ONE HAND IN A POCKET TO HELP PREVENT AN ELECTRICAL DISCHARGE PATH THROUGH THEIR BODY.

3. LARGE FILTER AND ENERGY STORAGE CAPACITORS CAN STORE A DANGEROUS CHARGE FOR SEVERAL DAYS OR MORE! NEVER TOUCH THE TERMINALS OF SUCH CAPACITORS! CAPACITORS CAN BE DISCHARGED BY CAREFULLY TOUCHING THE METAL TIP OF A SCREWDRIVER WITH AN INSULATED HANDLE ACROSS THEIR TERMINALS SEVERAL TIMES.

4. CHILDREN AND THOSE INEXPERIENCED IN WORKING WITH ELECTRONIC CIRCUITS SHOULD NOT ATTEMPT TO SERVICE LINE-POWERED CIRCUITS!

5. NEVER PLAY WITH ELECTRICITY!

6. AFTER SERVICING LINE-POWERED EQUIPMENT, REPLACE ALL PANELS AND SCREWS BEFORE APPLYING POWER.

7. WEAR RUBBER-SOLED SHOES AND STAND ON A DRY RUBBER MAT OR WOOD SURFACE WHEN WORKING WITH LINE-POWERED CIRCUITS.

# III. BASIC SEMICONDUCTOR CIRCUITS

## OVERVIEW

IN THIS ERA OF INTEGRATED CIRCUIT MICROCHIPS, THE SIMPLICITY AND ECONOMY OF CIRCUITS MADE FROM INDIVIDUAL COMPONENTS ARE OFTEN OVERLOOKED. THE CIRCUITS IN THIS SECTION SHOW MORE THAN 75 APPLICATIONS FOR SUCH BASIC COMPONENTS AS DIODES, TRANSISTORS, SCRs AND TRIACS. THESE CIRCUITS ARE PRECEDED BY BASIC RESISTOR AND CAPACITOR CIRCUITS THAT ARE ESSENTIAL INGREDIENTS IN MOST SEMICONDUCTOR CIRCUITS. SLIGHT VARIATIONS IN COMPONENTS AND CIRCUIT ASSEMBLY METHODS MAY CAUSE YOUR RESULTS TO DIFFER FROM THOSE DESCRIBED HERE. SINCE THE AUTHOR AND RADIOSHACK HAVE NO CONTROL OVER THE USE OF CIRCUITS YOU BUILD, WE ASSUME NO LIABILITY FOR SUCH USE.

## CIRCUIT ASSEMBLY TIPS

YOU CAN ASSEMBLE TEST VERSIONS OF CIRCUITS ON SOLDERLESS BREADBOARDS. AFTER YOU TEST AND EXPERIMENT WITH A CIRCUIT, YOU CAN ASSEMBLE A PERMANENT VERSION ON A CIRCUIT BOARD AND INSTALL IT IN AN ENCLOSURE. THOUGH EACH CIRCUIT THAT FOLLOWS INCLUDES SPECIFIC COMPONENT VALUES, SUBSTITUTIONS ARE USUALLY OK IF VOLTAGE, CURRENT AND POWER RATINGS ARE OBSERVED. FOR EXAMPLE, A 1.2K (1,200 OHMS) RESISTOR CAN USUALLY BE SUBSTITUTED FOR A 1K (1,000 OHMS) RESISTOR. A 100K (100,000 OHMS) POTENTIOMETER CAN BE USED IN PLACE OF A 50K (50,000 OHMS) POT. MANY NPN TRANSISTORS CAN BE SUBSTITUTED FOR THE POPULAR 2N2222.

# RESISTORS

RESISTORS RESIST THE FLOW OF AN ELECTRICAL CURRENT. THE UNIT OF RESISTANCE IS THE OHM ($\Omega$). A POTENTIAL DIFFERENCE OF ONE VOLT WILL FORCE A CURRENT OF ONE AMPERE THROUGH A RESISTANCE OF ONE OHM.

# OHM'S LAW

VOLTAGE (V) IS THE POTENTIAL DIFFERENCE ACROSS A RESISTOR. CURRENT (I) IS THE FLOW OF ELECTRONS THROUGH A RESISTOR. GIVEN ANY TWO VALUES OF RESISTANCE, VOLTAGE, OR CURRENT, THE THIRD VALUE CAN BE CALCULATED FROM OHM'S LAW:

$V = I \times R$ $\quad$ $I = V/R$ $\quad$ $R = V/I$

THE POWER DISSIPATED IN A RESISTOR CAN ALSO BE CALCULATED:

$P = V \times I$ $\quad$ $P = I^2 R$

THE UNIT OF POWER IS THE WATT. IT IS IMPORTANT TO BE SURE THAT ALL VALUES ARE EXPRESSED PROPERLY WHEN USING OHM'S LAW. FOR EXAMPLE, 65 MILLIVOLTS SHOULD BE EXPRESSED AS 0.065 VOLTS. 470 MILLIWATTS SHOULD BE EXPRESSED AS 0.47 WATTS. A 47K RESISTOR HAS A RESISTANCE OF 47 * 1,000 OR 47,000 OHMS. A 2.2M RESISTOR HAS A RESISTANCE OF 2.2 * 1,000,000 OR 2,200,000 OHMS.

USUALLY YOU MAY USE A RESISTOR WITH A VALUE WITHIN 10-20% OF THE REQUIRED VALUE. ALWAYS USE RESISTORS HAVING THE PROPER POWER RATING.

# RESISTORS IN SERIES

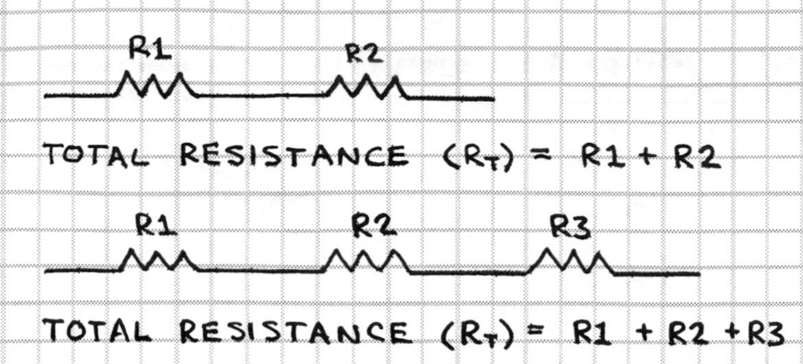


TOTAL RESISTANCE ($R_T$) = R1 + R2

TOTAL RESISTANCE ($R_T$) = R1 + R2 + R3

# RESISTORS IN PARALLEL

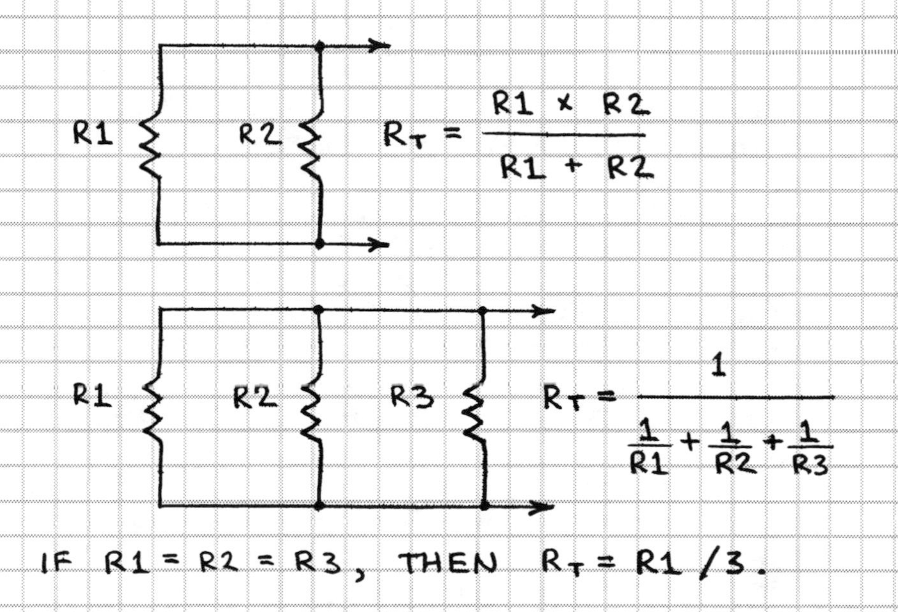


$$R_T = \frac{R1 \times R2}{R1 + R2}$$

$$R_T = \frac{1}{\frac{1}{R1} + \frac{1}{R2} + \frac{1}{R3}}$$

IF R1 = R2 = R3, THEN $R_T$ = R1 / 3.

# RESISTORS IN SERIES/PARALLEL

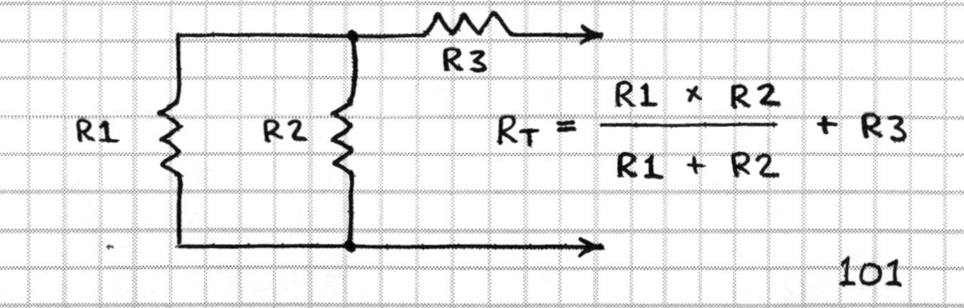


$$R_T = \frac{R1 \times R2}{R1 + R2} + R3$$

# HOW TO USE RESISTORS

## CURRENT LIMITING

A RESISTOR CAN BE PLACED IN SERIES WITH A LAMP, LED, SPEAKER, TRANSISTOR, OR OTHER COMPONENT TO REDUCE THE FLOW OF CURRENT THROUGH THE DEVICE. FOR EXAMPLE:

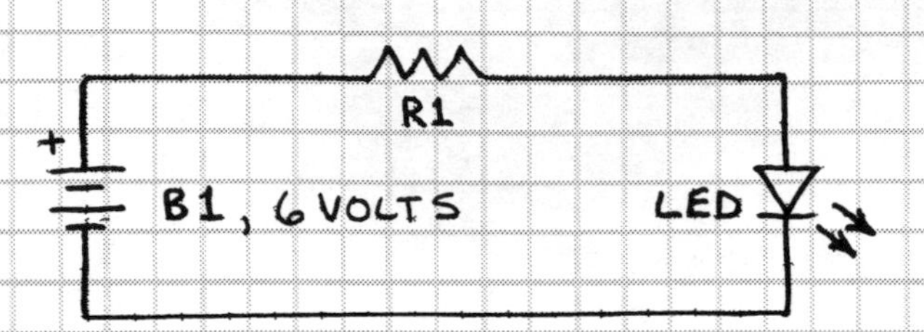


OHM'S LAW CAN BE USED TO CALCULATE THE CURRENT THROUGH THE LED FOR A RANGE OF STANDARD RESISTANCE VALUES. THE FORMULA FOR CURRENT IS $I = V/R$. AN LED DOES NOT BEGIN TO CONDUCT UNTIL THE FORWARD VOLTAGE IS ABOUT 1.7 VOLTS (RED LED). THEREFORE, THE FORMULA FOR CURRENT IS $I = (6-1.7)/R$.

| R1 (OHMS | LED CURRENT (AMPS) |
|---|---|
| 100 | .043 |
| 150 | .029 |
| 220 | .020 |
| 270 | .016 |
| 330 | .013 |

## VOLTAGE DIVISION

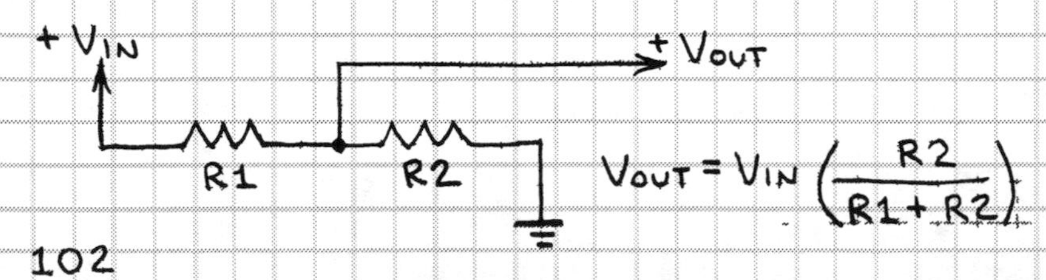

# WHEATSTONE BRIDGE

THE WHEATSTONE BRIDGE PERMITS VERY ACCURATE MEASUREMENTS OF RESISTANCE. HERE IS THE BASIC CIRCUIT:

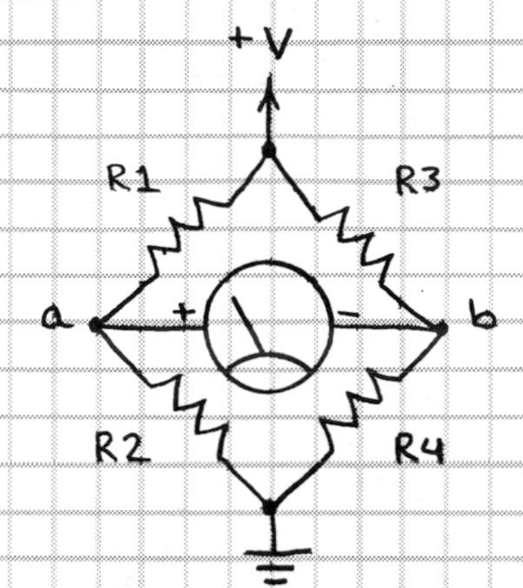


R1–R2 AND R3–R4 FORM TWO VOLTAGE DIVIDERS. WHEN THE VOLTAGE AT a EQUALS THE VOLTAGE AT b, THE METER INDICATES NO VOLTAGE AND THE BRIDGE IS SAID TO BE BALANCED. WHEN THIS OCCURS, THEN: R1/R3 = R2/R4.

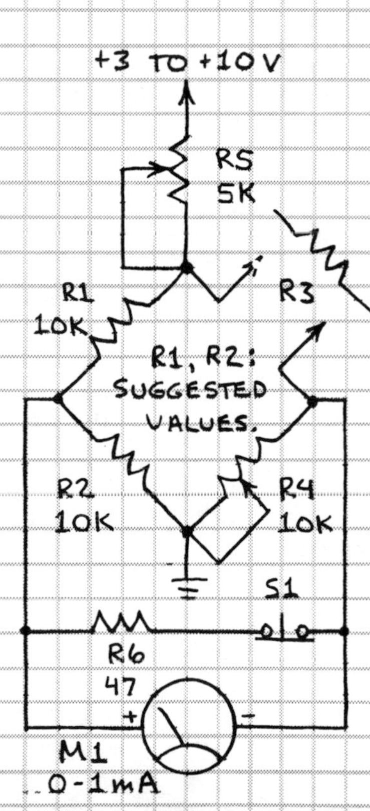


THE BRIDGE SHOWN HERE PERMITS THE ACCURATE MEASUREMENT OF AN UNKNOWN RESISTANCE (R3). R1 AND R2 SHOULD BE PRECISION (1%) RESISTORS. R4 IS A POTENTIOMETER WITH A CALIBRATED DIAL. R5 IS USED TO REGULATE THE CURRENT FROM THE POWER SUPPLY. R6 AND S1 FORM A SHUNT THAT PROTECTS M1. ADJUST R4 UNTIL M1 = 0. PRESS S1 AND REPEAT. R3 = R4. IF R1 ≠ R2, THEN R3 = (R1 × R4) / R2.

# CAPACITORS

CAPACITORS STORE AN ELECTRICAL CHARGE. THE UNIT OF CAPACITANCE IS THE <u>FARAD</u>. A 1-FARAD CAPACITOR CONNECTED TO A 1-VOLT SUPPLY WILL STORE A CHARGE OF $6.28 \times 10^{18}$ ELECTRONS. MOST CAPACITORS HAVE CONSIDERABLY LESS CAPACITY. VALUES COMMONLY RANGE FROM A FEW PICOFARADS ($10^{-12}$ FARAD) TO A FEW THOUSAND MICROFARADS ($10^{-6}$ FARAD).

1 FARAD = 1 F
1 MICROFARAD = 1 $\mu$F = $10^{-6}$ F
1 NANOFARAD = 1 nF = $10^{-9}$ F
1 PICOFARAD = 1 pF = $10^{-12}$ F

A CAPACITOR CAN BE CHARGED ALMOST INSTANTLY BY CONNECTING ITS LEADS DIRECTLY ACROSS A POWER SUPPLY. THE CHARGING TIME CAN BE INCREASED BY INSERTING A RESISTOR BETWEEN THE SUPPLY AND THE CAPACITOR.

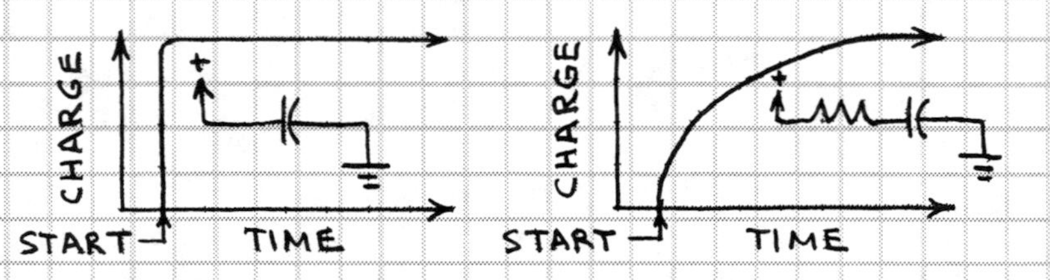


A CHARGED CAPACITOR WILL GRADUALLY LOSE ITS CHARGE THROUGH LEAKAGE. THE DISCHARGE TIME CAN BE REDUCED BY CONNECTING A RESISTOR ACROSS THE CAPACITOR'S TWO LEADS:

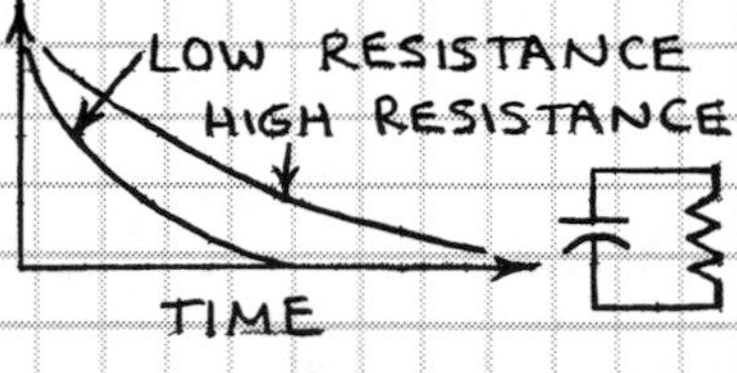

# CAPACITORS IN SERIES

C1 C2

TOTAL CAPACITANCE ($C_T$) = $\frac{C1 \times C2}{C1 + C2}$

C1 C2 C3

TOTAL CAPACITANCE ($C_T$) = $\frac{1}{\frac{1}{C1}+\frac{1}{C2}+\frac{1}{C3}}$

# CAPACITORS IN PARALLEL

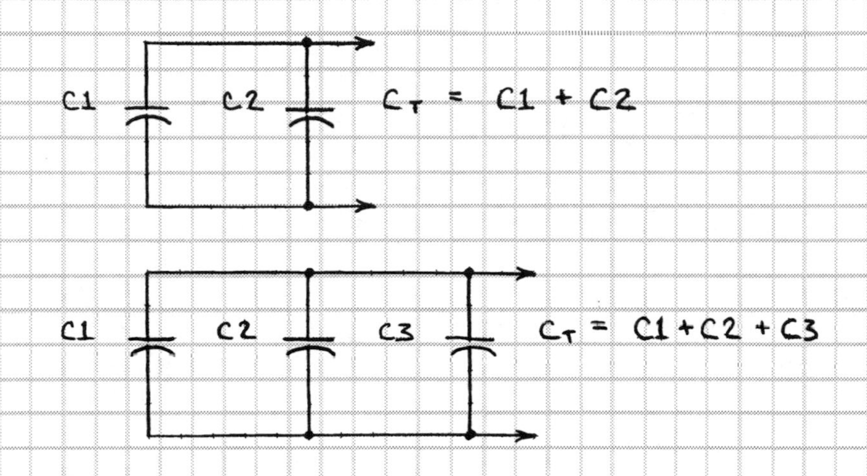


## WARNING!

MOST CAPACITORS CAN RETAIN A CHARGE FOR A CONSIDERABLE TIME AFTER THE CHARGING SUPPLY HAS BEEN SWITCHED OFF. THEREFORE USE CAUTION WHEN WORKING WITH CAPACITORS. A LARGE ELECTROLYTIC CAPACITOR CHARGED TO ONLY 5 TO 10 VOLTS CAN MELT THE TIP OF A SCREWDRIVER SHORTED ACROSS ITS LEADS! HIGH-VOLTAGE CAPACITORS IN TV SETS AND PHOTOFLASH UNITS CAN STORE A LETHAL CHARGE!

# HOW TO USE CAPACITORS

## SIGNAL FILTERING

A SINGLE CAPACITOR CAN DIVERT AN UNWANTED SIGNAL TO GROUND:

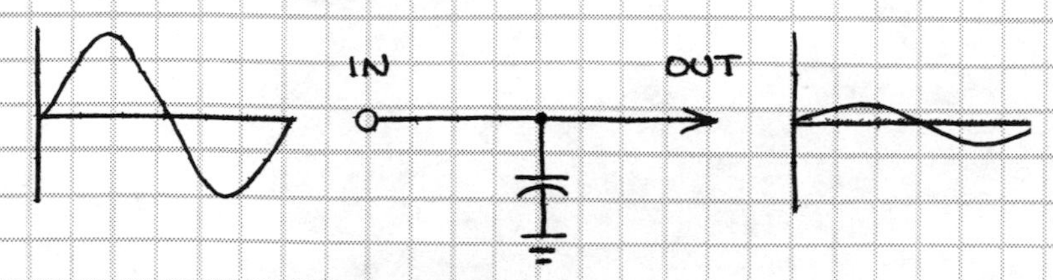


A SINGLE CAPACITOR CAN REMOVE AN UNWANTED DC COMPONENT FROM A FLUCTUATING SIGNAL:

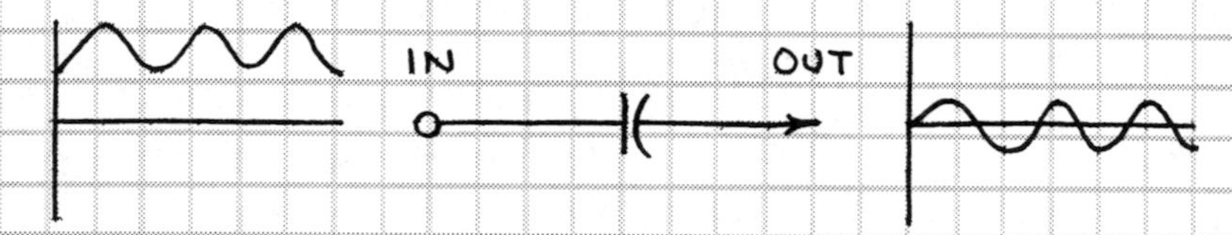


## POWER SUPPLY FILTERING

A LARGE CAPACITOR WILL SMOOTH THE PULSATING VOLTAGE FROM A POWER SUPPLY INTO STEADY DIRECT CURRENT:

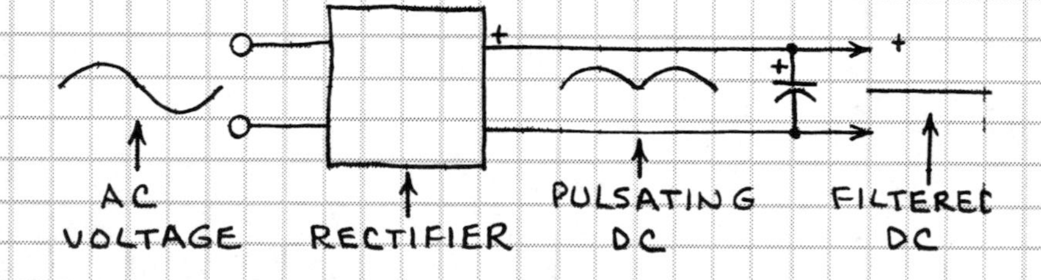


## SPIKE AND NOISE SUPPRESSION

A 0.1 μF CAPACITOR ACROSS THE POWER SUPPLY PINS OF A LOGIC CHIP WILL HELP SUPPRESS FALSE TRIGGERING CAUSED BY BRIEF POWER SUPPLY NOISE SPIKES.

# RESISTOR-CAPACITOR CIRCUITS

AMONG THE MOST IMPORTANT OF ALL CIRCUITS ARE THE BASIC RESISTOR-CAPACITOR (RC) CIRCUITS:

## INTEGRATOR

THE INTEGRATOR IS AN RC CIRCUIT THAT TRANSFORMS AN INCOMING SQUARE WAVE INTO A TRIANGLE WAVE:

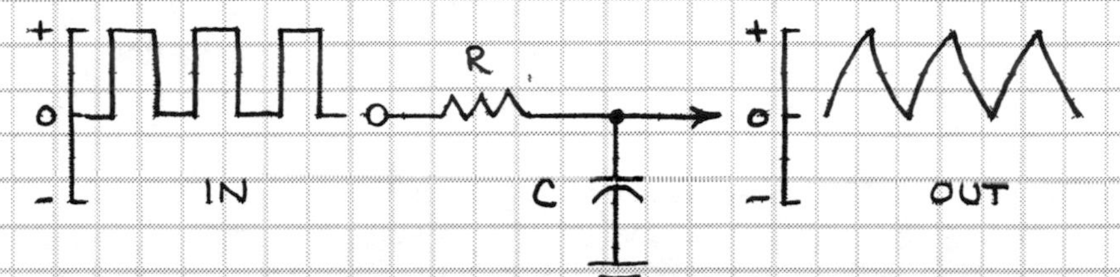


R*C IS THE TIME CONSTANT OF THE CIRCUIT. RC MUST BE AT LEAST 10 TIMES THE PERIOD OF THE INPUT SIGNAL. IF NOT, THE AMPLITUDE OF THE OUTPUT SIGNAL WILL BE REDUCED. THE CIRCUIT WILL THEN BE A LOW-PASS FILTER THAT BLOCKS HIGH FREQUENCIES.

## DIFFERENTIATOR

THE DIFFERENTIATOR IS AN RC CIRCUIT THAT TRANSFORMS AN INCOMING SQUARE WAVE INTO A PULSED OR SPIKED WAVEFORM:

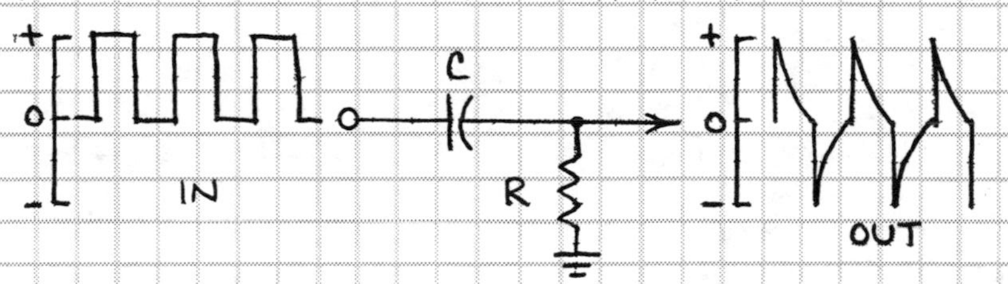


THE RC TIME CONSTANT SHOULD BE 1/10 (OR LESS) OF THE DURATION OF THE INCOMING PULSES. DIFFERENTIATORS ARE OFTEN USED TO CREATE TRIGGER PULSES.

# DIODES AND RECTIFIERS

DIODES AND RECTIFIERS ARE SEMICONDUCTOR DEVICES THAT CONDUCT ELECTRICITY IN ONLY ONE DIRECTION. IT IS IMPORTANT TO UNDERSTAND THAT A DIODE DOES NOT BEGIN TO CONDUCT UNTIL THE FORWARD VOLTAGE REACHES A THRESHOLD POINT. FOR SILICON DIODES THIS VOLTAGE IS ABOUT 0.6 VOLT. FOR GERMANIUM DIODES IT IS ABOUT 0.3 VOLT. THIS GRAPH SUMS UP DIODE OPERATION:

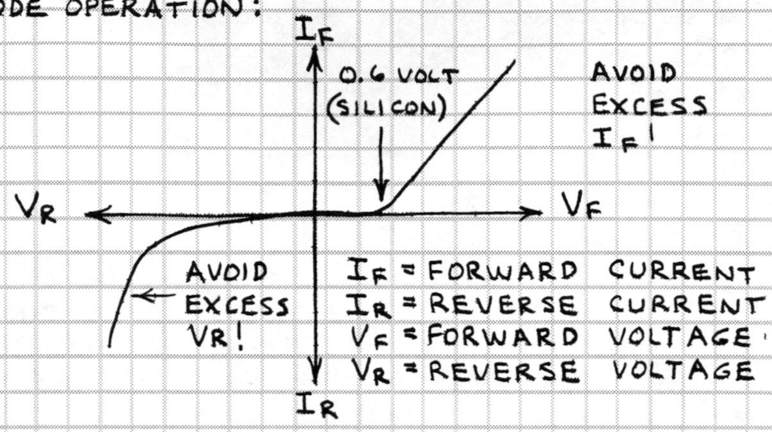


# VOLTAGE DROPPER

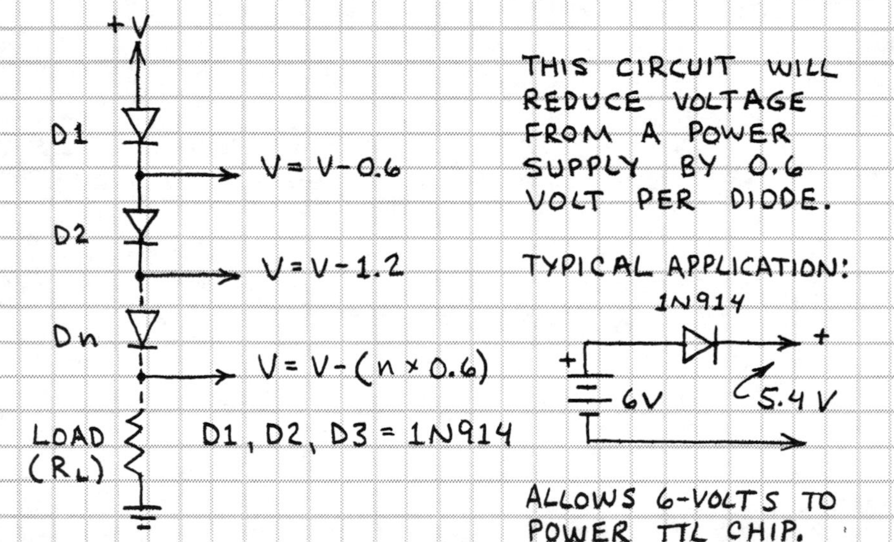


THIS CIRCUIT WILL REDUCE VOLTAGE FROM A POWER SUPPLY BY 0.6 VOLT PER DIODE.

TYPICAL APPLICATION:

ALLOWS 6-VOLTS TO POWER TTL CHIP.

# VOLTAGE REGULATOR

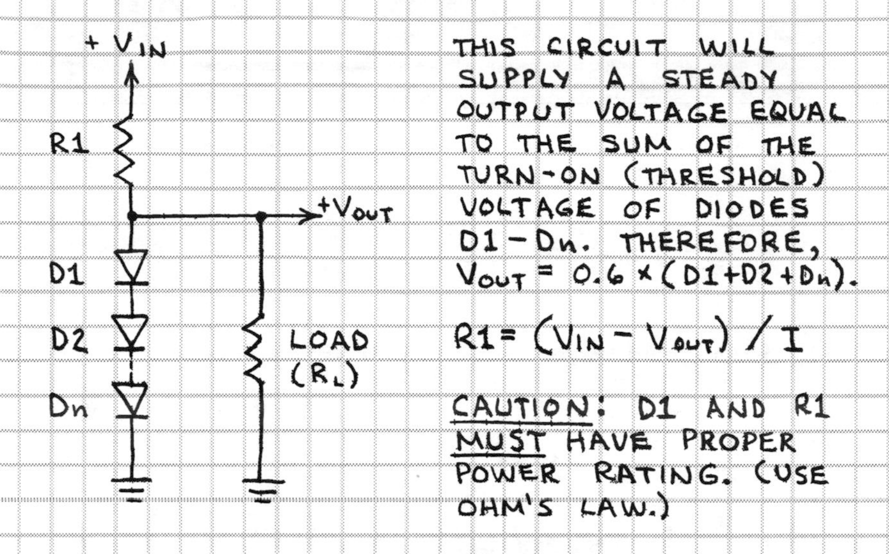


THIS CIRCUIT WILL SUPPLY A STEADY OUTPUT VOLTAGE EQUAL TO THE SUM OF THE TURN-ON (THRESHOLD) VOLTAGE OF DIODES D1 – Dn. THEREFORE, $V_{OUT} = 0.6 \times (D1 + D2 + Dn)$.

$R1 = (V_{IN} - V_{OUT}) / I$

<u>CAUTION</u>: D1 AND R1 <u>MUST</u> HAVE PROPER POWER RATING. (USE OHM'S LAW.)

# TRIANGLE-TO-SINE WAVE

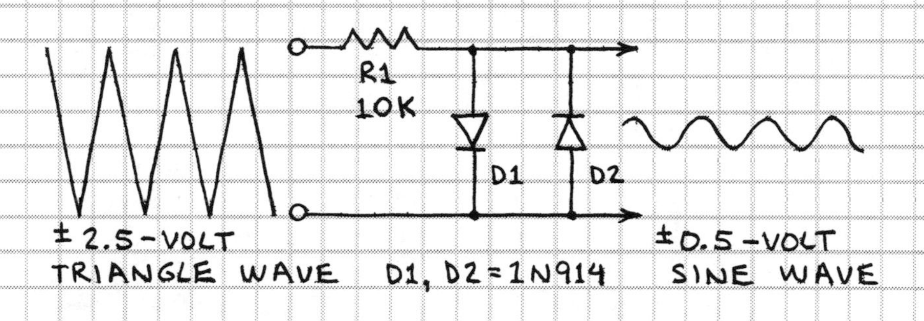


± 2.5-VOLT TRIANGLE WAVE

D1, D2 = 1N914

± 0.5-VOLT SINE WAVE

# PEAK-READING VOLTMETER

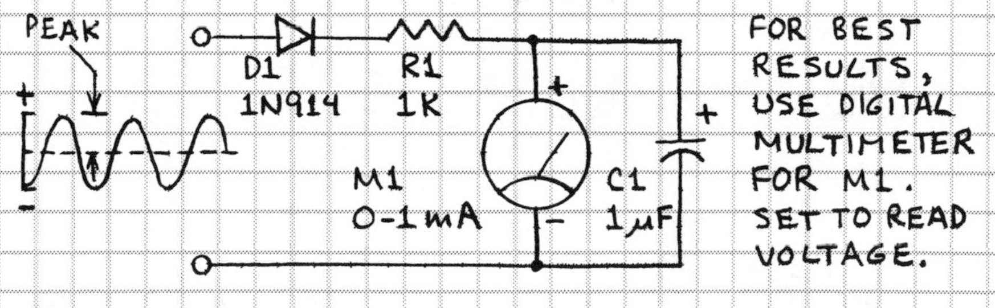


FOR BEST RESULTS, USE DIGITAL MULTIMETER FOR M1. SET TO READ VOLTAGE.

FREQUENCY OF INCOMING SIGNAL MUST BE HIGH ENOUGH TO KEEP C1 CHARGED.

# REVERSE-POLARITY PROTECTOR

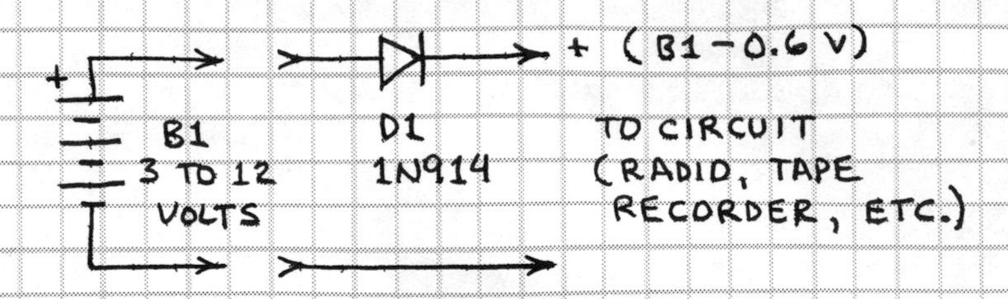


DIODE PROTECTS CIRCUIT IF BATTERY IS INSTALLED WITH REVERSED POLARITY.

# TRANSIENT PROTECTOR

WHEN THE CURRENT FLOWING THROUGH AN INDUCTOR IS SUDDENLY SWITCHED OFF, THE COLLAPSING MAGNETIC FIELD WILL GENERATE A HIGH VOLTAGE IN THE INDUCTOR'S COILS. THIS VOLTAGE SPIKE MAY HAVE AN AMPLITUDE OF HUNDREDS OR EVEN THOUSANDS OF VOLTS. A DIODE CAN PROTECT THE CIRCUIT TO WHICH THE INDUCTOR IS CONNECTED BY PROVIDING A SHORT CIRCUIT FOR THE HIGH VOLTAGE SPIKE. FOR EXAMPLE:

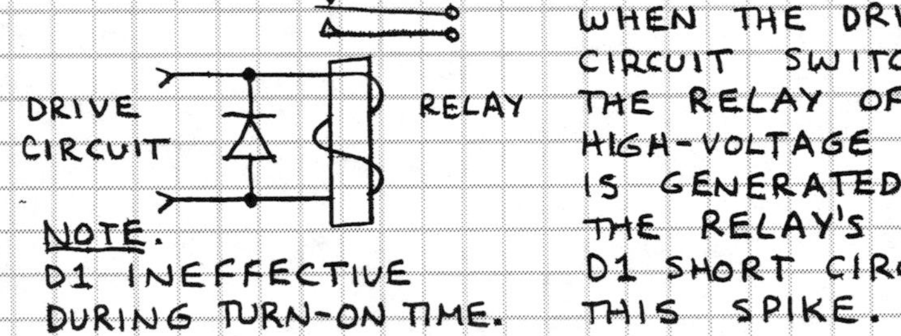


NOTE.
D1 INEFFECTIVE DURING TURN-ON TIME.

WHEN THE DRIVE CIRCUIT SWITCHES THE RELAY OFF, A HIGH-VOLTAGE SPIKE IS GENERATED IN THE RELAY'S COIL. D1 SHORT CIRCUITS THIS SPIKE.

# METER PROTECTOR

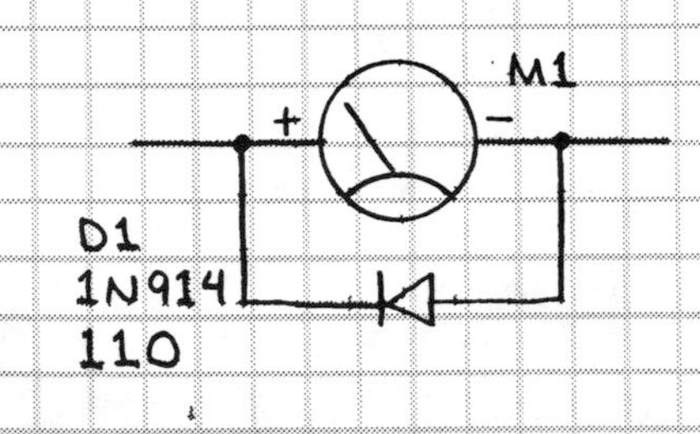


CONNECT A DIODE ACROSS THE TERMINALS OF A METER TO PROVIDE REVERSE CURRENT PROTECTION.

# ADJUSTABLE WAVEFORM CLIPPER

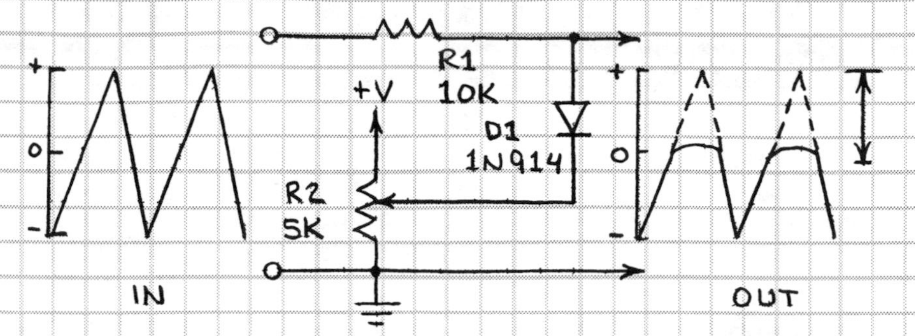


ADJUST R2 TO CONTROL CLIPPING AMPLITUDE. +V SHOULD BE A VOLT OR SO HIGHER THAN PEAK INPUT VOLTAGE.

# ADJUSTABLE ATTENUATOR

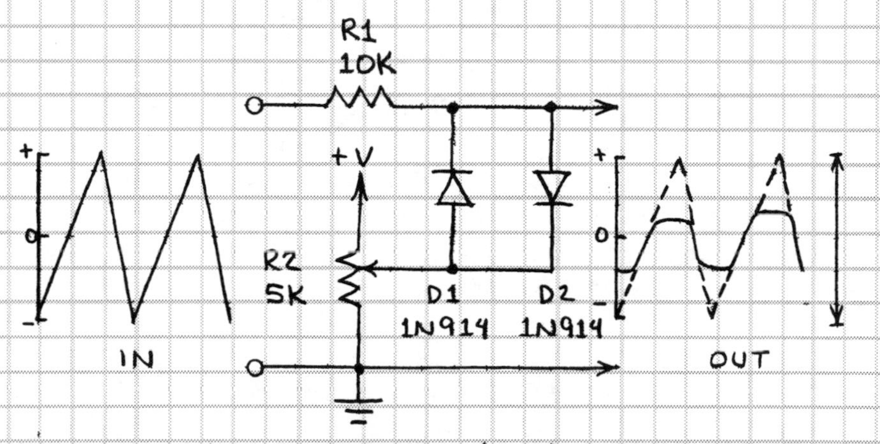


THIS IS A BIPOLARITY (+/-) VERSION OF THE ADJUSTABLE CLIPPER.

# AUDIO LIMITER

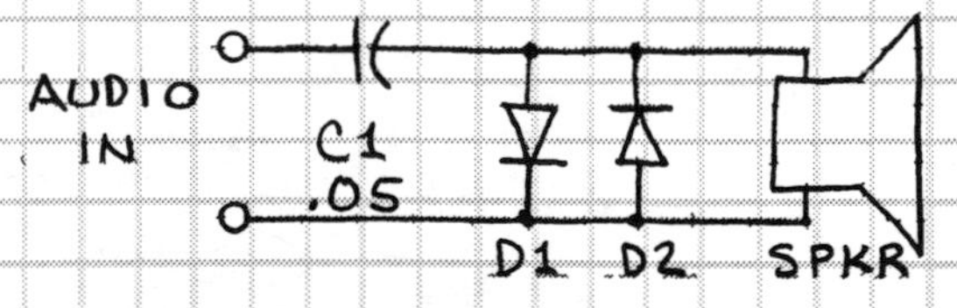


USE TO LIMIT NOISE, POPS, AND STATIC.

D1, D2 = 1N914

# HALF-WAVE RECTIFIER

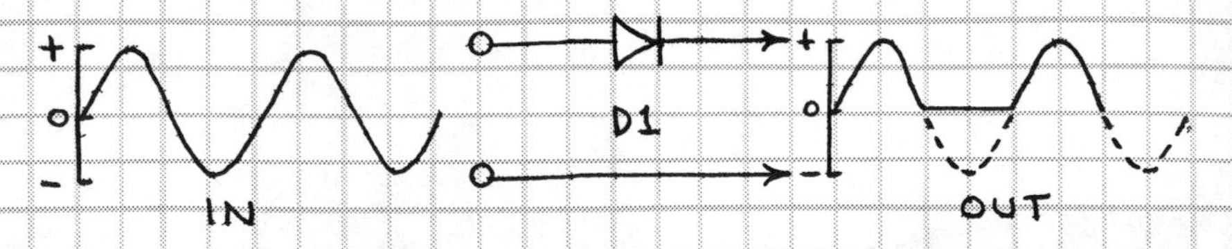


D1 IS ANY DIODE RATED FOR THE INPUT VOLTAGE. THIS CIRCUIT IS USED TO TRANSFORM AN AC WAVE INTO PULSATING DC AND TO DETECT MODULATED RADIO SIGNALS.

# DUAL HALF-WAVE RECTIFIER

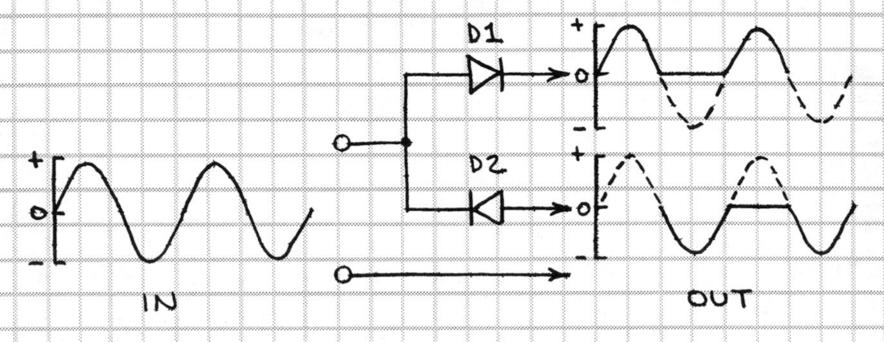


THIS CIRCUIT TRANSFORMS BOTH HALVES OF AN AC WAVE INTO PULSATING DC.

# FULL-WAVE RECTIFIER

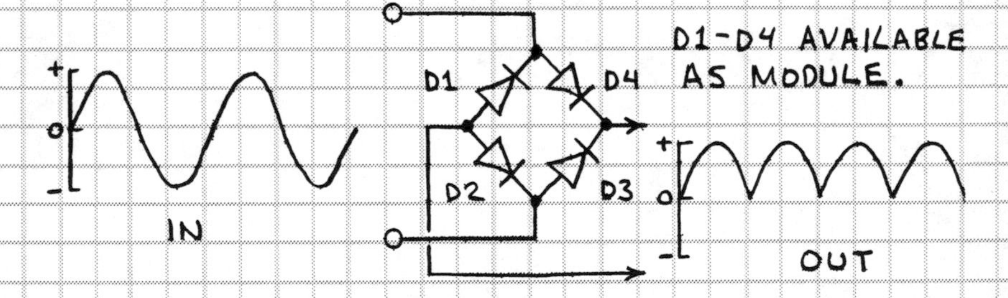


ALSO CALLED A BRIDGE RECTIFIER. USED TO TRANSFORM BOTH HALVES OF AC WAVE TO DC.

# CASCADE VOLTAGE DOUBLER

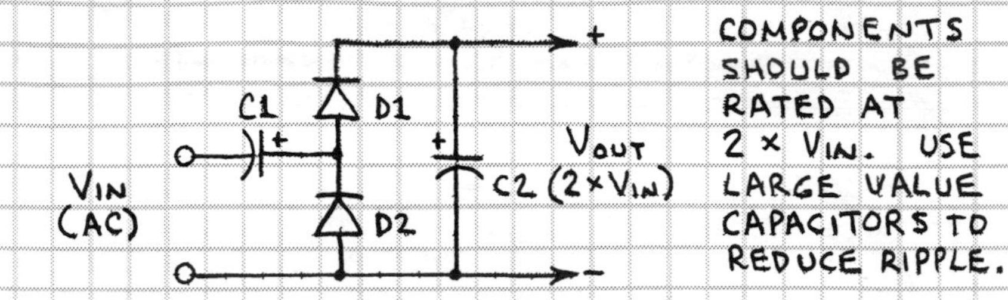


COMPONENTS SHOULD BE RATED AT 2 × VIN. USE LARGE VALUE CAPACITORS TO REDUCE RIPPLE.

# BRIDGE VOLTAGE DOUBLER

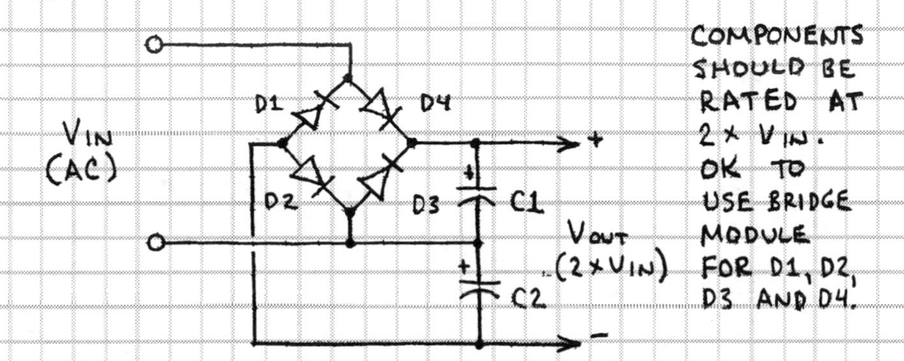


COMPONENTS SHOULD BE RATED AT 2 × VIN. OK TO USE BRIDGE MODULE FOR D1, D2, D3 AND D4.

# VOLTAGE QUADRUPLER

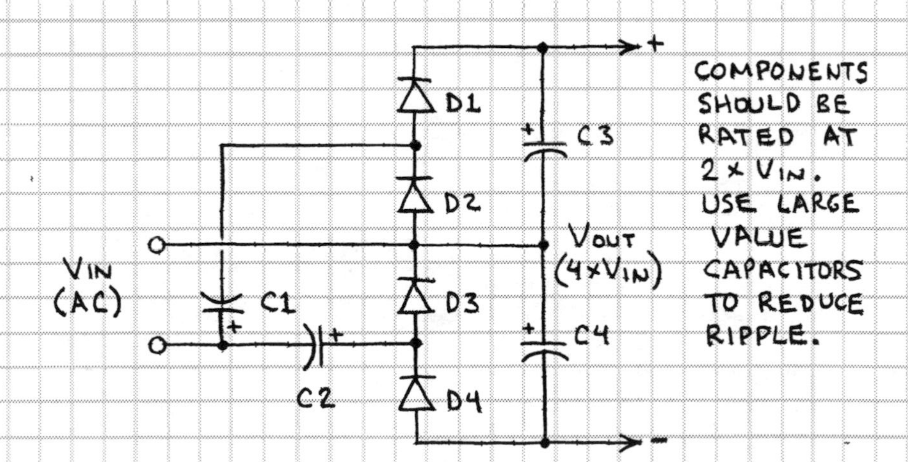


COMPONENTS SHOULD BE RATED AT 2 × VIN. USE LARGE VALUE CAPACITORS TO REDUCE RIPPLE.

CAUTION: VOLTAGE MULTIPLICATION CIRCUITS CAN PRODUCE HIGH VOLTAGES. USE CARE!

# DIODE LOGIC GATES

THESE SIMPLE LOGIC CIRCUITS CAN BE USED TO TEACH BASICS OF DIGITAL LOGIC AND IN PRACTICAL APPLICATIONS.

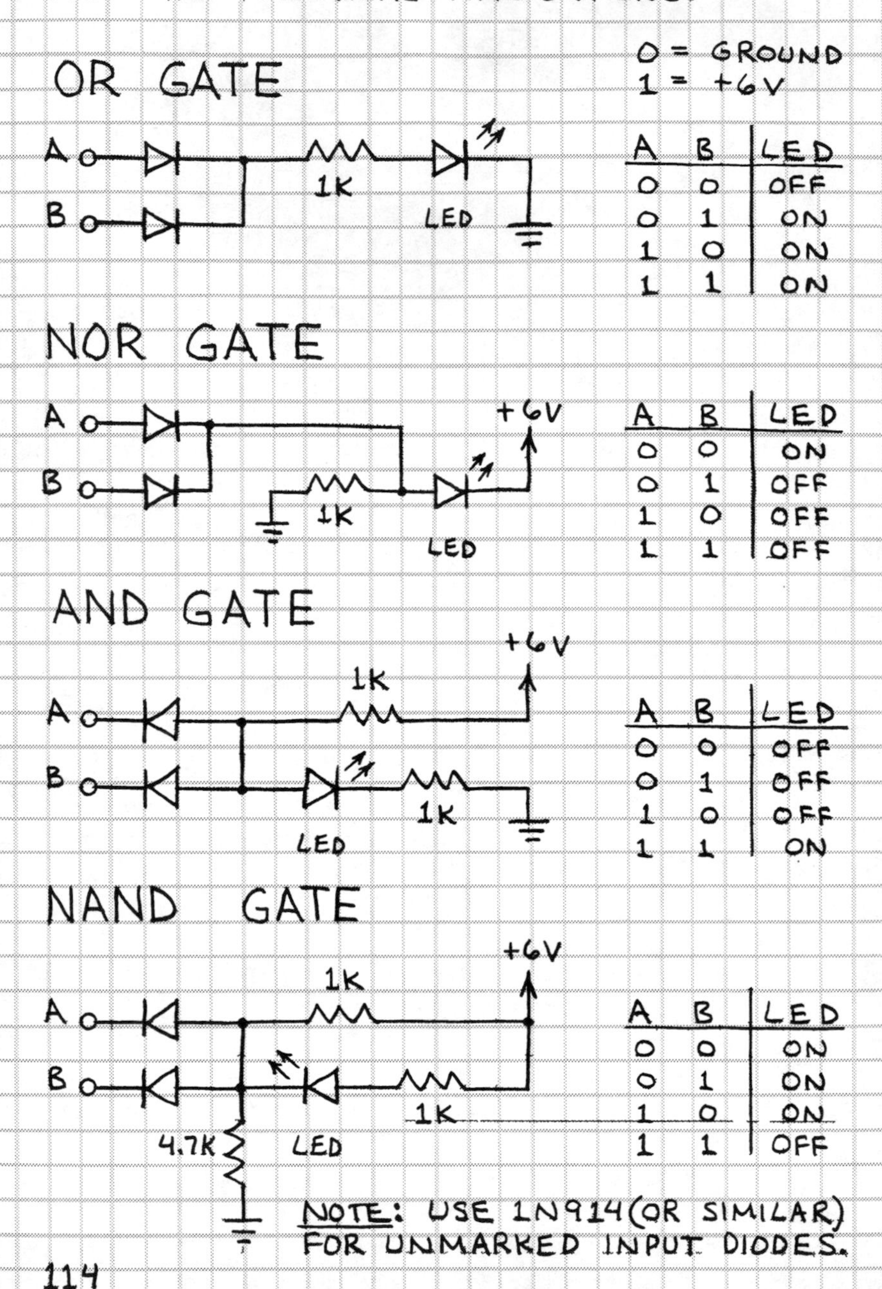


## OR GATE

| A | B | LED |
|---|---|---|
| 0 | 0 | OFF |
| 0 | 1 | ON |
| 1 | 0 | ON |
| 1 | 1 | ON |

## NOR GATE

| A | B | LED |
|---|---|---|
| 0 | 0 | ON |
| 0 | 1 | OFF |
| 1 | 0 | OFF |
| 1 | 1 | OFF |

## AND GATE

| A | B | LED |
|---|---|---|
| 0 | 0 | OFF |
| 0 | 1 | OFF |
| 1 | 0 | OFF |
| 1 | 1 | ON |

## NAND GATE

| A | B | LED |
|---|---|---|
| 0 | 0 | ON |
| 0 | 1 | ON |
| 1 | 0 | ON |
| 1 | 1 | OFF |

NOTE: USE 1N914 (OR SIMILAR) FOR UNMARKED INPUT DIODES.

# DECIMAL-TO-BINARY ENCODER

THIS CIRCUIT IS A PROGRAMMABLE READ-ONLY MEMORY (PROM). USE 1N914 DIODES.

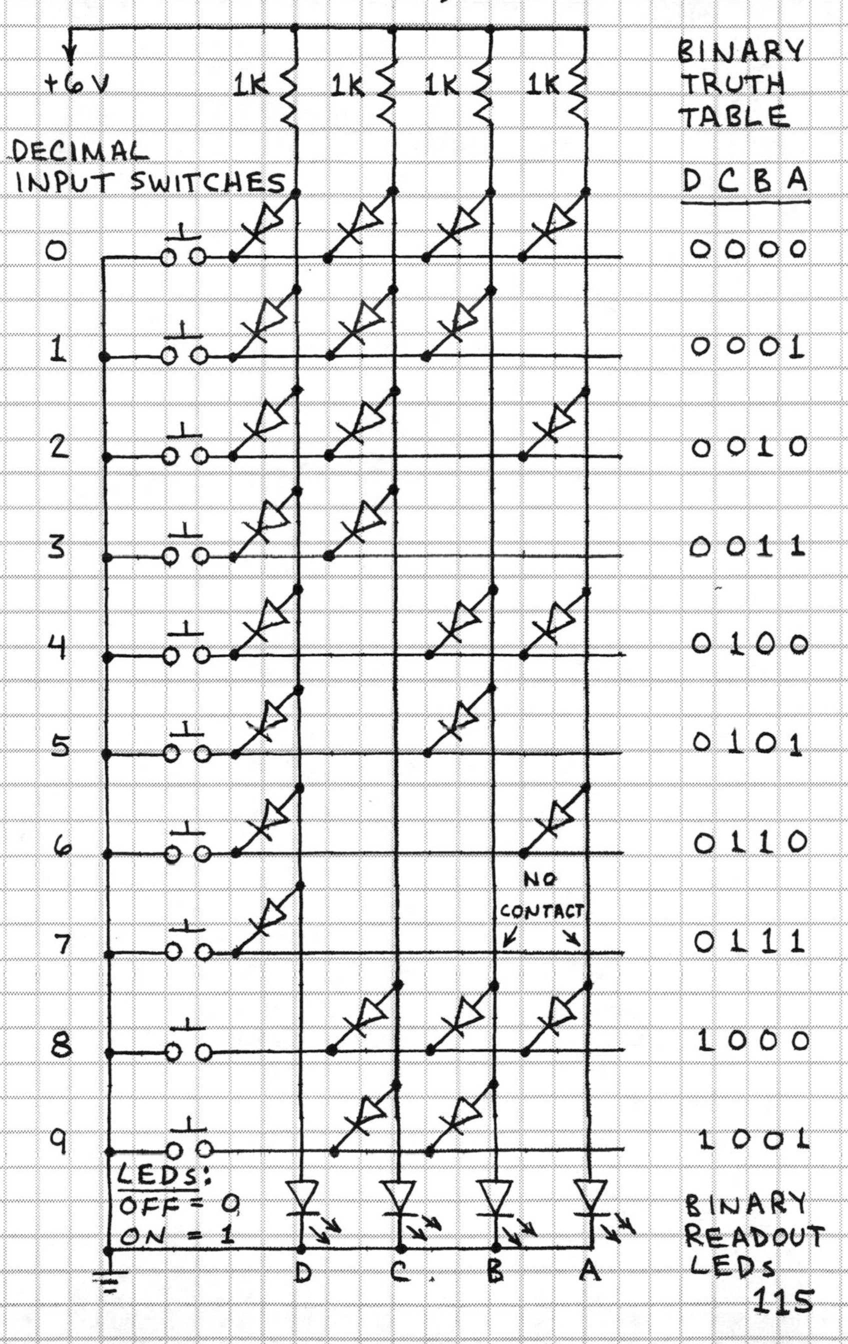

# ZENER DIODES

NORMALLY A CURRENT DOES NOT FLOW THROUGH A DIODE CONNECTED IN THE REVERSE DIRECTION. THE ZENER DIODE IS DESIGNED SPECIFICALLY TO BEGIN CONDUCTING IN THE REVERSE DIRECTION WHEN THE REVERSE VOLTAGE EXCEEDS A THRESHOLD VALUE (THE BREAKDOWN VOLTAGE). THEREFORE THE ZENER DIODE IS A VOLTAGE-SENSITIVE SWITCH. THIS GRAPH SUMS UP ZENER DIODE OPERATION:

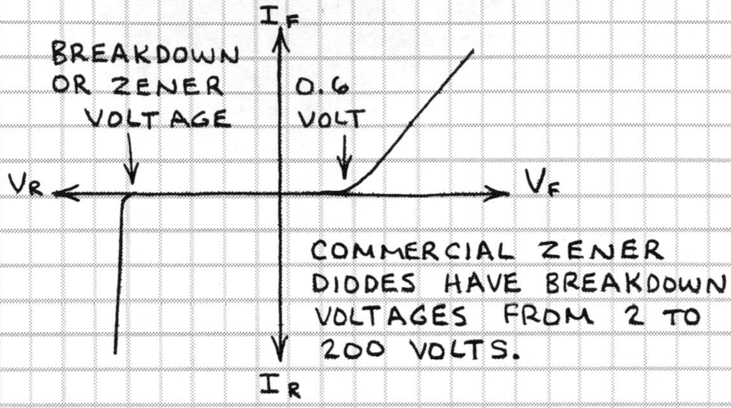


COMMERCIAL ZENER DIODES HAVE BREAKDOWN VOLTAGES FROM 2 TO 200 VOLTS.

# VOLTAGE REGULATOR MODEL

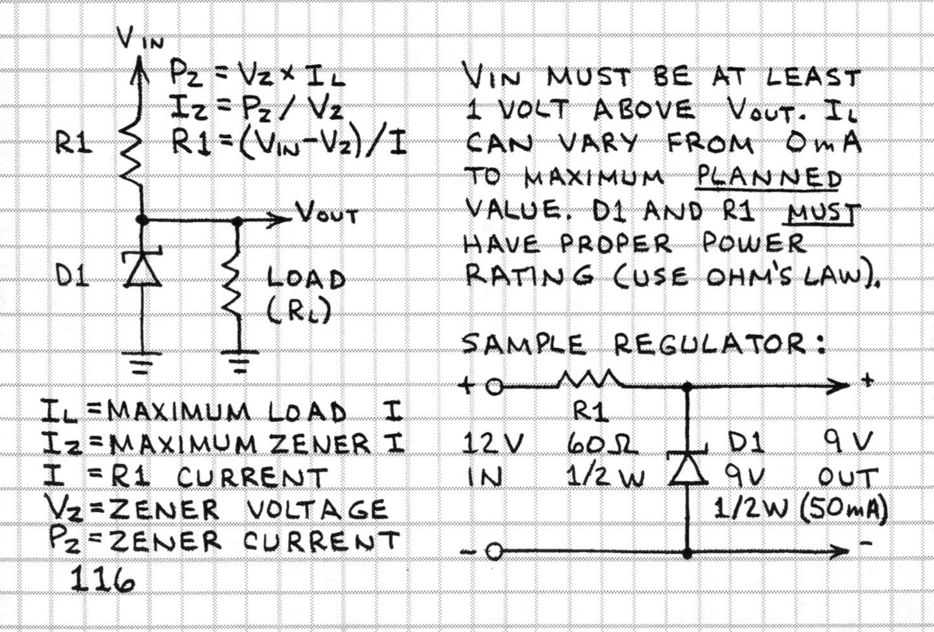


$V_{IN}$ MUST BE AT LEAST 1 VOLT ABOVE $V_{OUT}$. $I_L$ CAN VARY FROM 0mA TO MAXIMUM PLANNED VALUE. D1 AND R1 MUST HAVE PROPER POWER RATING (USE OHM'S LAW).

$I_L$ = MAXIMUM LOAD I
$I_Z$ = MAXIMUM ZENER I
$I$ = R1 CURRENT
$V_Z$ = ZENER VOLTAGE
$P_Z$ = ZENER CURRENT

# VOLTAGE INDICATOR

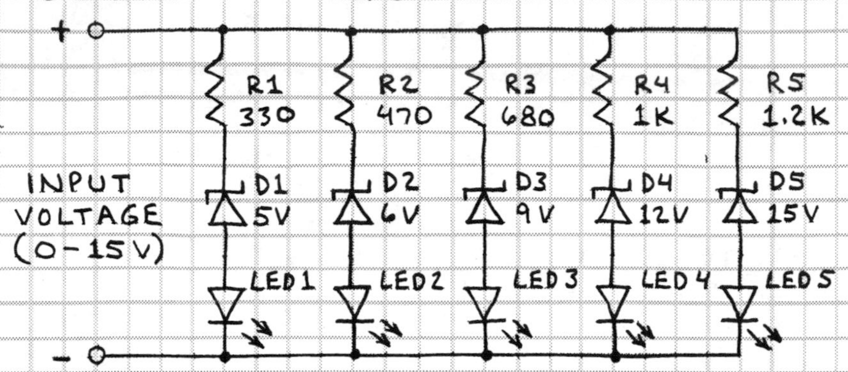


LEDs GLOW IN SEQUENCE AS INPUT VOLTAGE RISES. OK TO USE DIFFERENT ZENERS SO LONG AS SERIES RESISTOR LIMITS CURRENT THROUGH LED TO SAFE VALUE.

# VOLTAGE SHIFTER

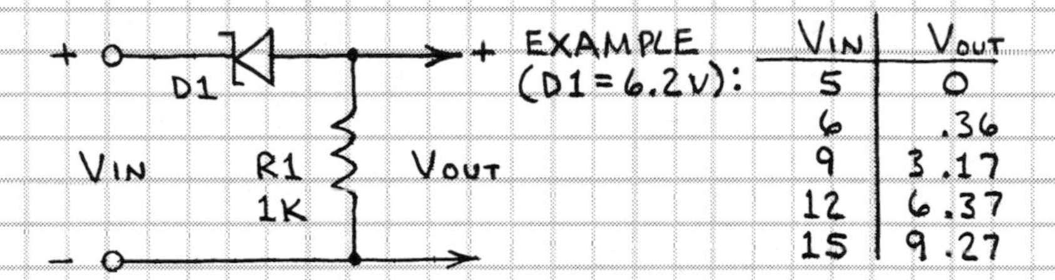


| $V_{IN}$ | $V_{OUT}$ |
|---|---|
| 5 | 0 |
| 6 | .36 |
| 9 | 3.17 |
| 12 | 6.37 |
| 15 | 9.27 |

# WAVEFORM CLIPPERS

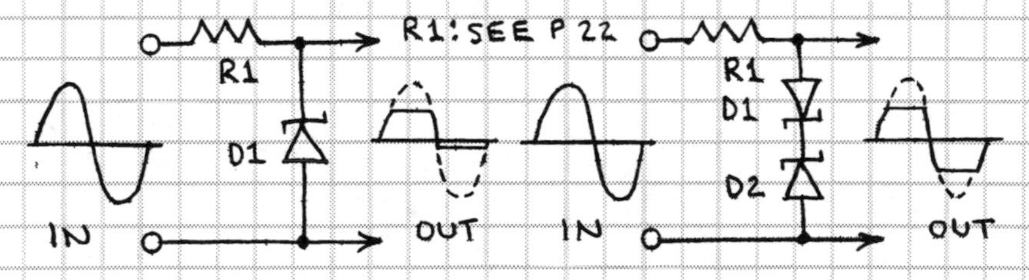


USE TO REDUCE LEVEL OF INCOMING SIGNAL. ALSO CONVERTS SINE WAVE TO NEAR SQUARE WAVE.

CLIPS **BOTH** HALVES OF WAVE (EQUALLY WHEN D1=D2). USE AS POP FILTER FOR SPEAKERS AND PHONES.

# BIPOLAR TRANSISTORS

A BIPOLAR TRANSISTOR IS A 3-TERMINAL SEMICONDUCTOR DEVICE IN WHICH A SMALL CURRENT AT ONE TERMINAL CAN CONTROL A MUCH LARGER CURRENT FLOWING BETWEEN THE SECOND AND THIRD TERMINAL. THIS MEANS TRANSISTORS CAN FUNCTION AS BOTH AMPLIFIERS AND SWITCHES. BIPOLAR TRANSISTORS ARE CLASSIFIED AS NPN OR PNP ACCORDING TO THE DOPING CONTAINED IN THEIR THREE REGIONS.

# BASIC TRANSISTOR SWITCHES

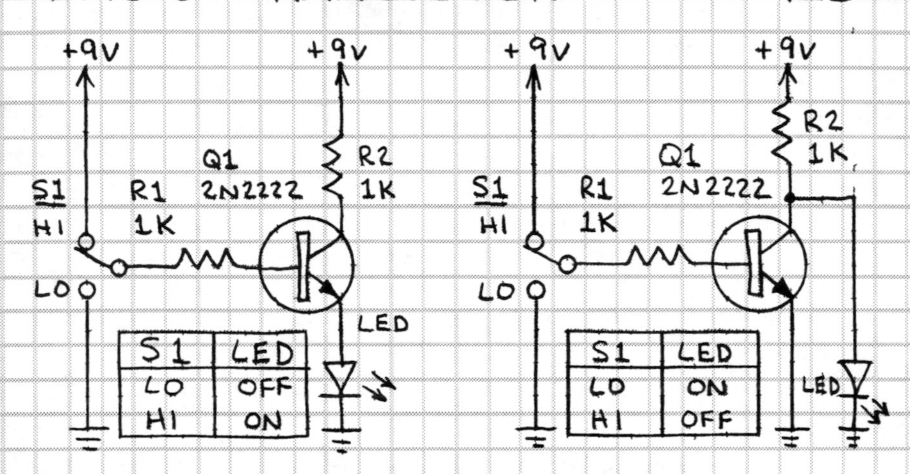


# BASIC TRANSISTOR AMPLIFIER

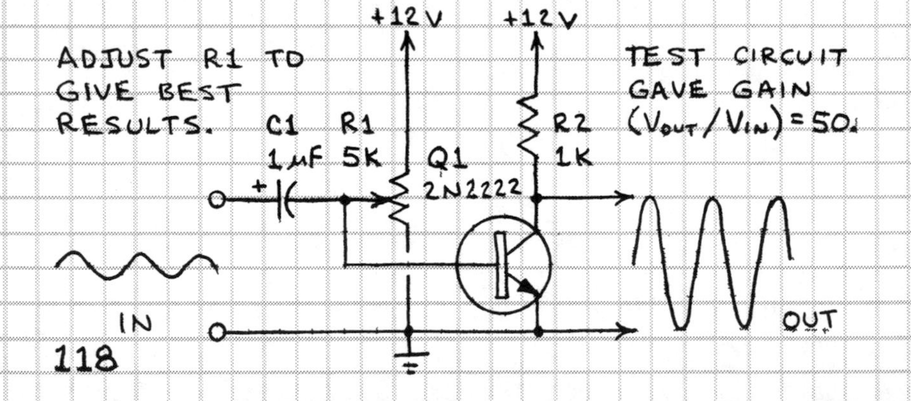

# RELAY DRIVER

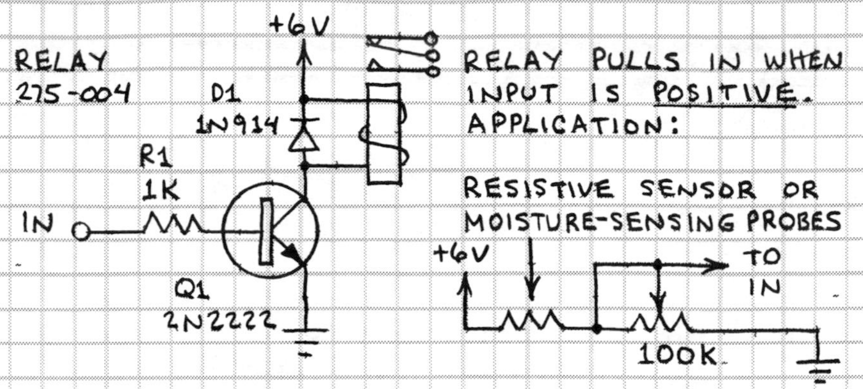


RELAY PULLS IN WHEN INPUT IS POSITIVE. APPLICATION:

# RELAY CONTROLLER

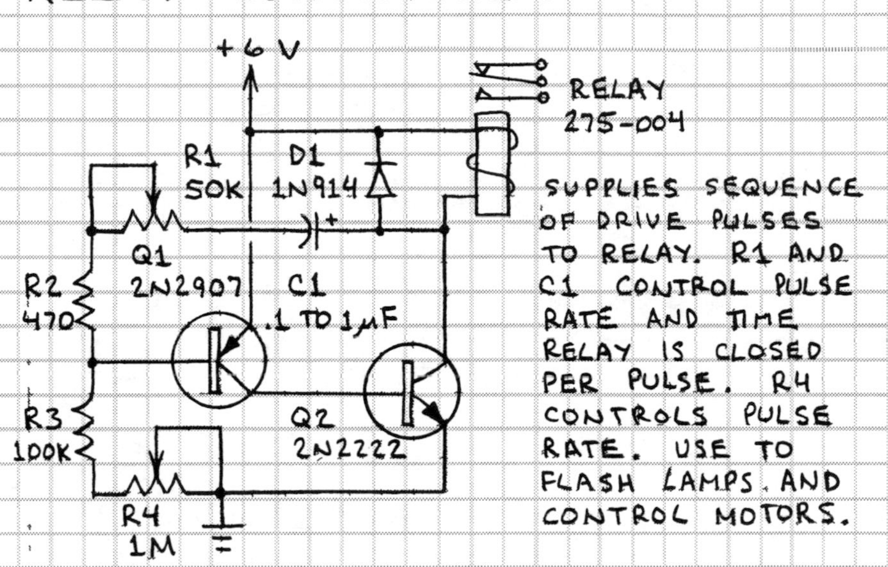


SUPPLIES SEQUENCE OF DRIVE PULSES TO RELAY. R1 AND C1 CONTROL PULSE RATE AND TIME RELAY IS CLOSED PER PULSE. R4 CONTROLS PULSE RATE. USE TO FLASH LAMPS AND CONTROL MOTORS.

# LED REGULATOR

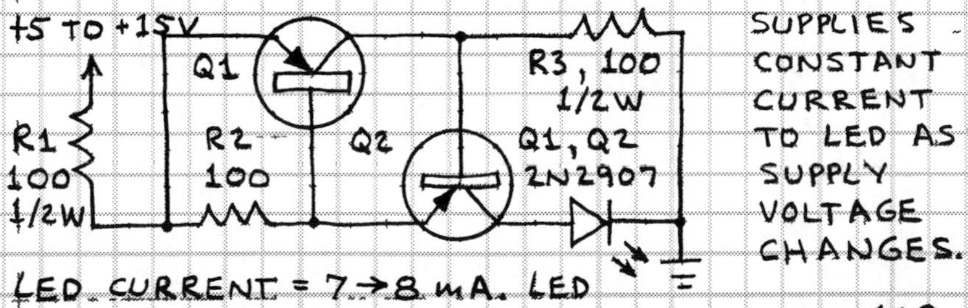


SUPPLIES CONSTANT CURRENT TO LED AS SUPPLY VOLTAGE CHANGES.

LED CURRENT = 7→8 mA.

# 3-VOLT SPEAKER AMPLIFIER

USE TO GIVE LOW-POWER SPEAKER TO RADIOS AND TAPE PLAYERS WITHOUT SPEAKERS.

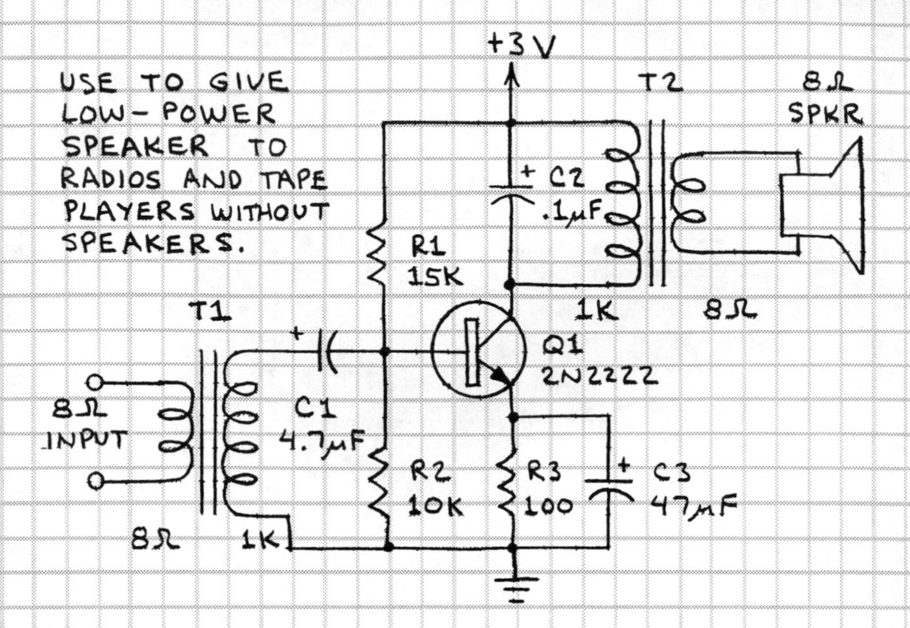


# 2-STAGE SPEAKER AMPLIFIER

THIS CIRCUIT REQUIRES NO INPUT TRANSFORMER.

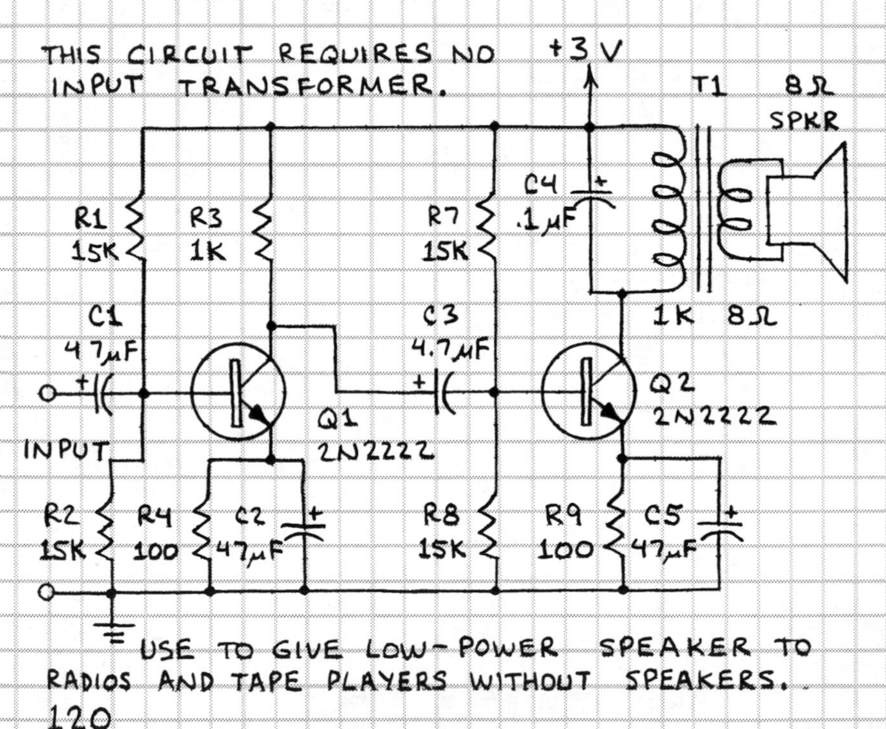


USE TO GIVE LOW-POWER SPEAKER TO RADIOS AND TAPE PLAYERS WITHOUT SPEAKERS.

# MICROPHONE PREAMPLIFIER

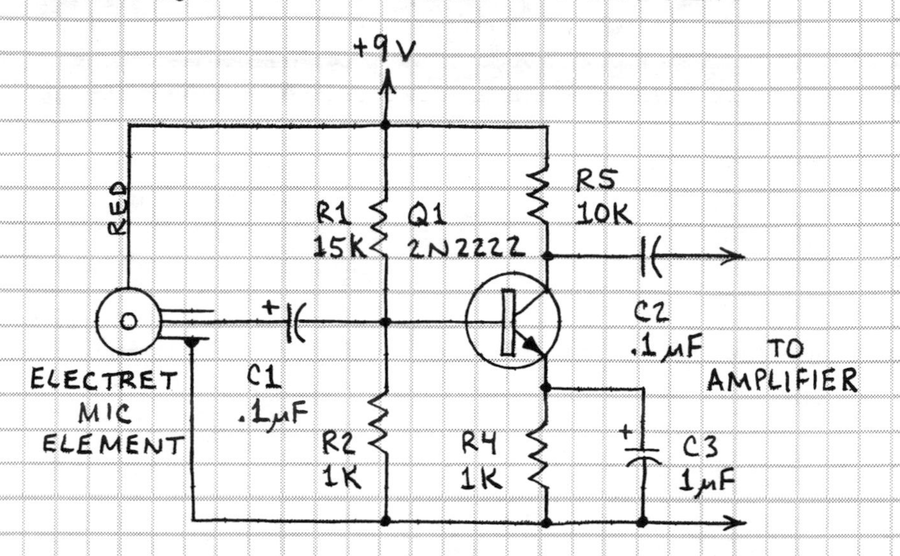


USE WITH TAPE RECORDERS, PUBLIC ADDRESS SYSTEMS AND PORTABLE AMPLIFIERS.

# AUDIO MIXER

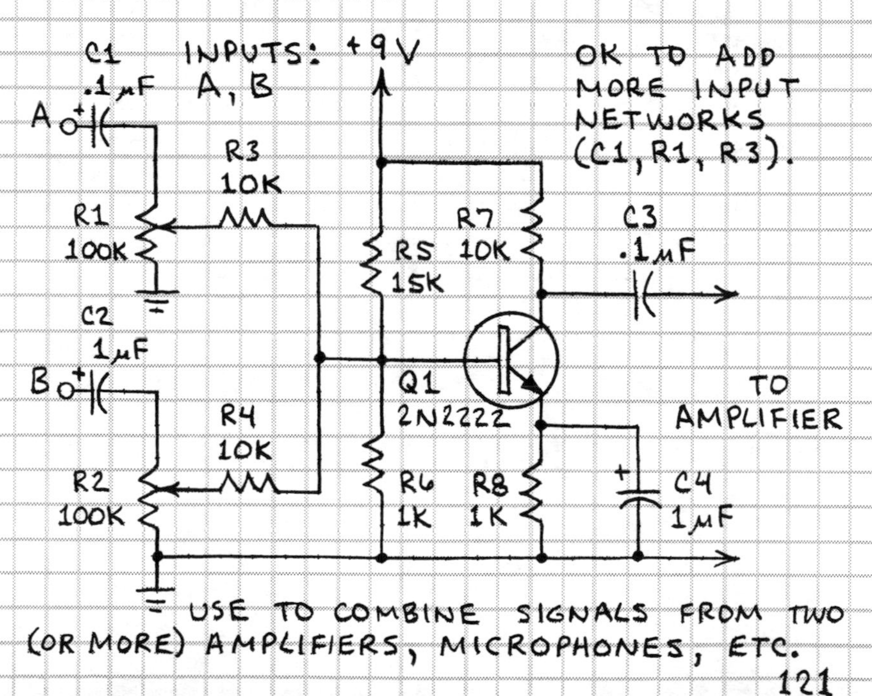


USE TO COMBINE SIGNALS FROM TWO (OR MORE) AMPLIFIERS, MICROPHONES, ETC.

# AUDIO OSCILLATOR

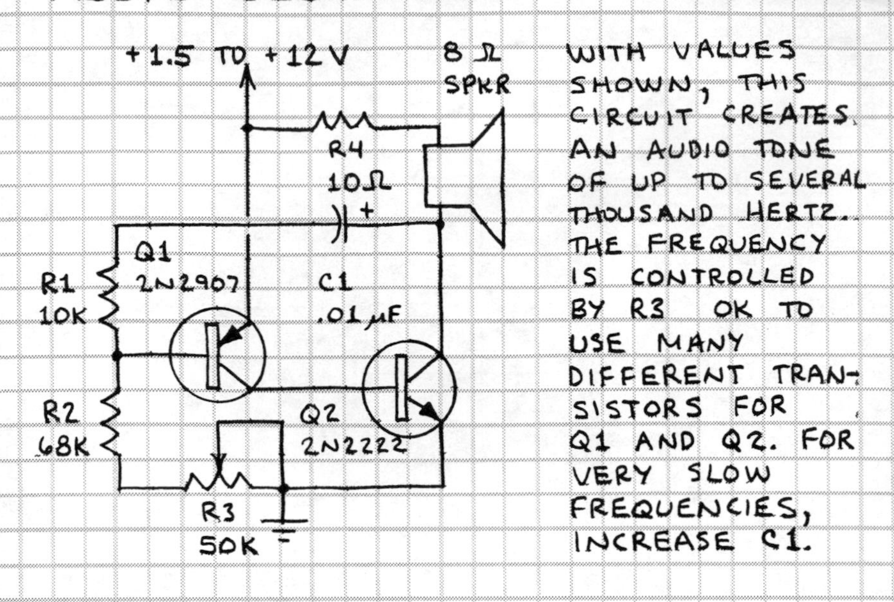


WITH VALUES SHOWN, THIS CIRCUIT CREATES AN AUDIO TONE OF UP TO SEVERAL THOUSAND HERTZ. THE FREQUENCY IS CONTROLLED BY R3. OK TO USE MANY DIFFERENT TRANSISTORS FOR Q1 AND Q2. FOR VERY SLOW FREQUENCIES, INCREASE C1.

# METRONOME

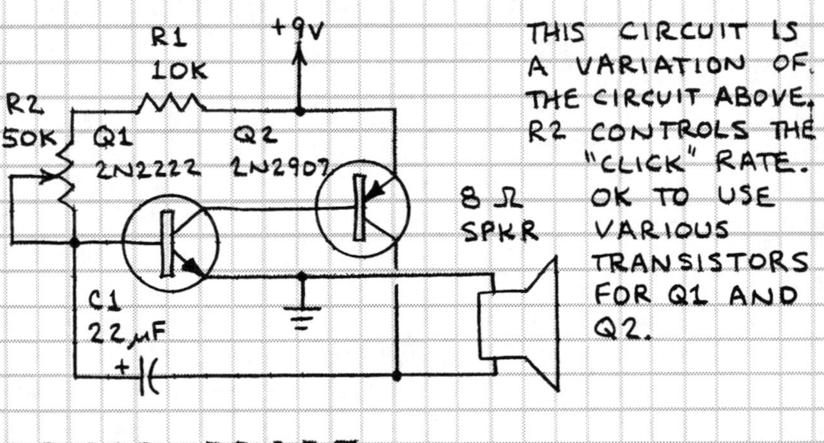


THIS CIRCUIT IS A VARIATION OF THE CIRCUIT ABOVE. R2 CONTROLS THE "CLICK" RATE. OK TO USE VARIOUS TRANSISTORS FOR Q1 AND Q2.

# LOGIC PROBE

TO LOGIC CIRCUIT

Q1

R1 10K

+5V

R2 1K

LED

Q1 2N2222

| LOGIC IN | LED |
|---|---|
| LO | OFF |
| HI | ON |

# ADJUSTABLE SIREN

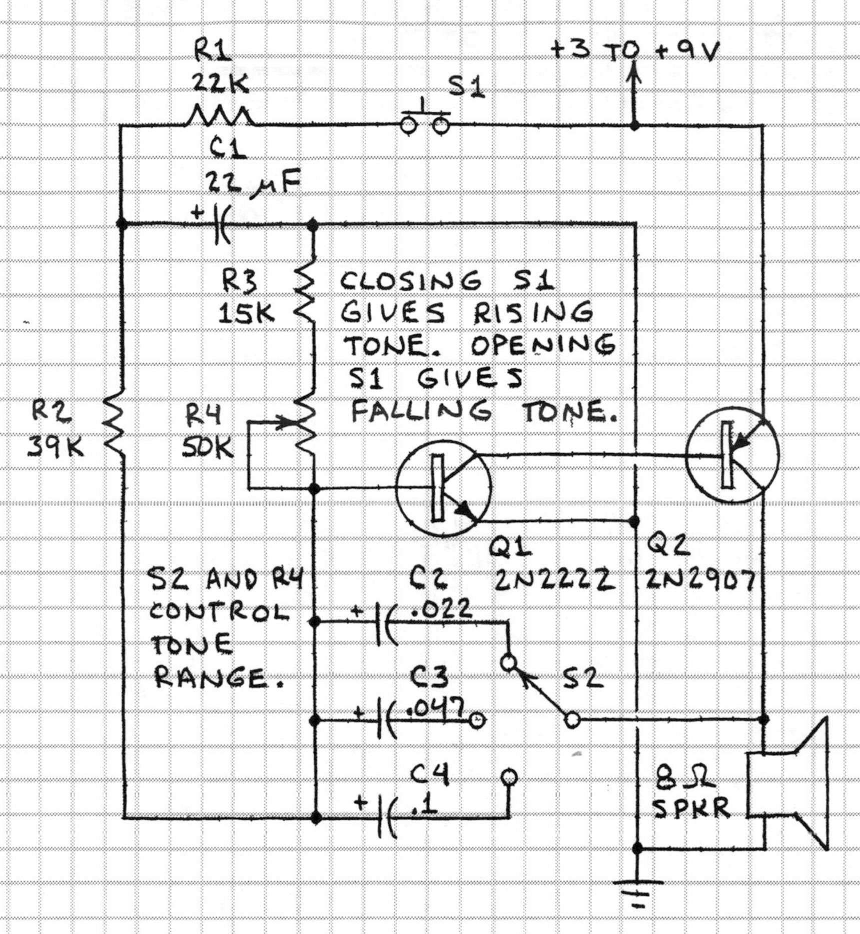


# AUDIO NOISE GENERATOR

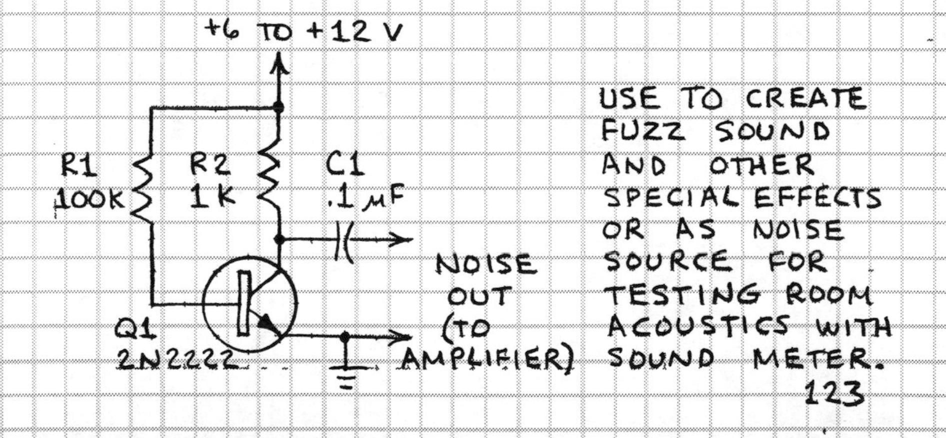


USE TO CREATE FUZZ SOUND AND OTHER SPECIAL EFFECTS OR AS NOISE SOURCE FOR TESTING ROOM ACOUSTICS WITH SOUND METER.

# 1-TRANSISTOR OSCILLATOR

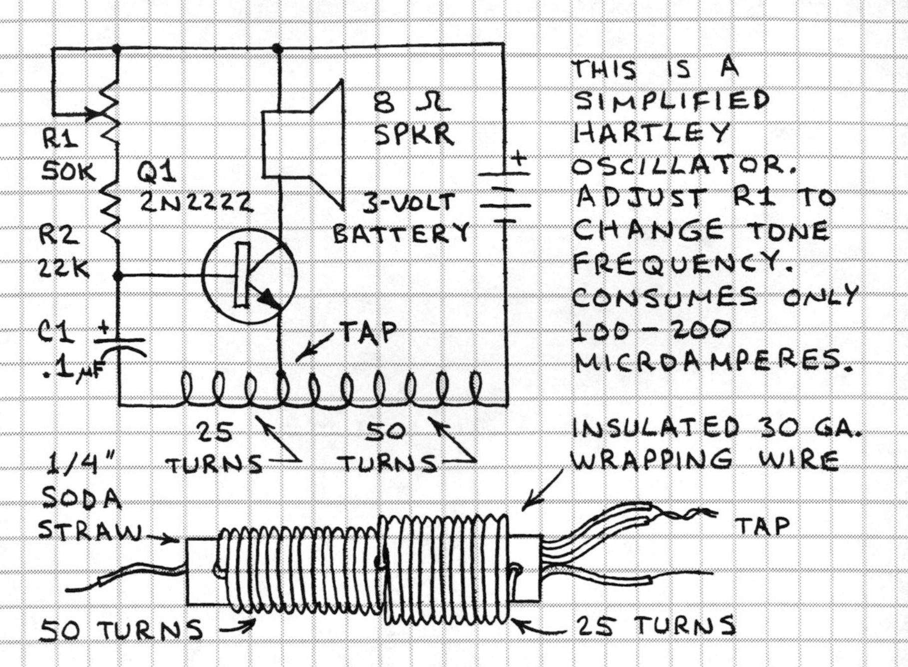


THIS IS A SIMPLIFIED HARTLEY OSCILLATOR. ADJUST R1 TO CHANGE TONE FREQUENCY. CONSUMES ONLY 100 – 200 MICROAMPERES.

COIL: PUNCH TWO SMALL HOLES 1-1/8" APART IN STRAW. INSERT WIRE IN FIRST HOLE, WIND 50 TURNS, INSERT WIRE LOOP IN SECOND HOLE, AND WIND BACK 25 TURNS. PUNCH HOLE THROUGH FIRST WINDING AND INSERT END OF WIRE. TAP: CUT LOOP AND TWIST EXPOSED WIRES

# SWITCH DEBOUNCER

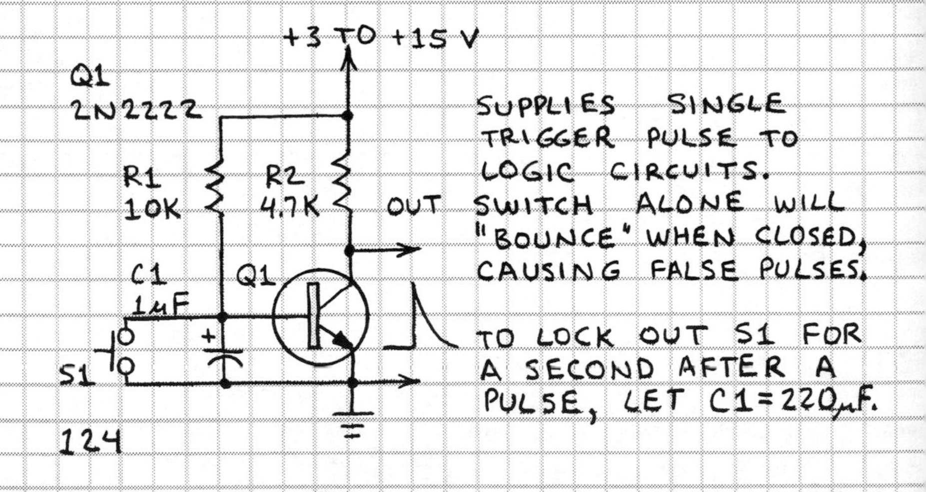


SUPPLIES SINGLE TRIGGER PULSE TO LOGIC CIRCUITS. SWITCH ALONE WILL "BOUNCE" WHEN CLOSED, CAUSING FALSE PULSES.

TO LOCK OUT S1 FOR A SECOND AFTER A PULSE, LET C1=220μF.

# MINIATURE RF TRANSMITTER

THIS CIRCUIT IS PATTERNED AFTER A PILL-SIZED BIOTELEMETRY TRANSMITTER FIRST DEVELOPED BY DR. R. STEWART MACKAY AND OTHER MEDICAL RESEARCHERS IN THE LATE 1950'S. THIS TRANSMITTER REMAINS ONE OF THE SMALLEST EVER DEVELOPED.

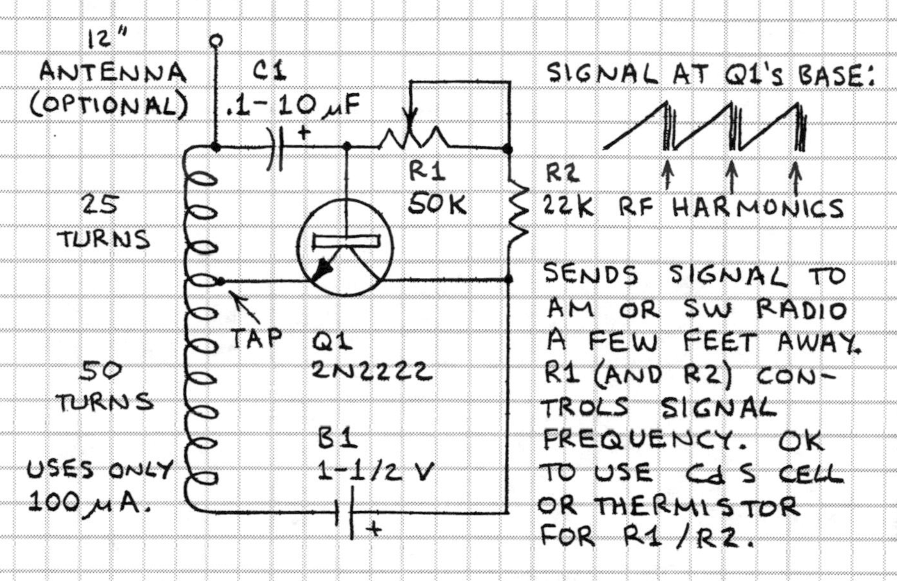


COIL: USE THE COIL SHOWN ON THE FACING PAGE OR MAKE A MUCH SMALLER VERSION WITH A 1/2" LENGTH OF SODA STRAW AND NO. 30 MAGNET WIRE. BURN THE VARNISH FROM THE LAST 1/4" OF THE COIL'S LEADS (USE A MATCH). THEN LIGHTLY BUFF THE CHARRED VARNISH WITH FINE SAND PAPER.

B1: USE A PENLIGHT CELL OR A MERCURY OR SILVER OXIDE BUTTON CELL. WARNING: NEVER ATTEMPT TO SOLDER LEADS TO MINIATURE POWER CELLS. THEY WILL EXPLODE.

C1: 0.1 µF GIVES AUDIO TONE; 10 µF GIVES AUDIBLE CLICKS. INSERT FERRITE CORE OR STEEL NAIL IN COIL TO ALTER THE SIGNAL. USE MINIATURE ELECTROLYTIC CAPACITOR.

# FREQUENCY METER

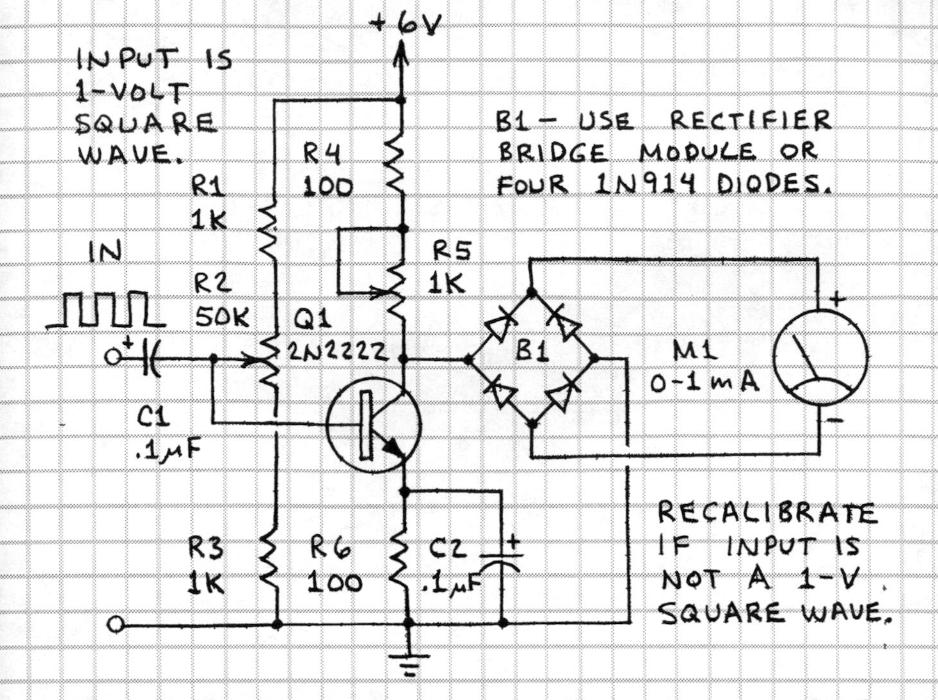


THIS CIRCUIT IS SUITABLE FOR SPECIFIC ROLES RATHER THAN GENERAL FREQUENCY MEASUREMENTS. TO CALIBRATE FOR 0-1 KHz RANGE:

1 SET R2 AND R5 AT MID POINTS.
2. APPLY 1 KHz, 1 VOLT SQUARE WAVE AT INPUT.
3. ADJUST R2 UNTIL M1 = 1 mA.
4. REMOVE 1 KHz SIGNAL.
5. ADJUST R3 UNTIL M1 = 0.
6. REAPPLY 1 KHz SIGNAL.
7. ADJUST R2 UNTIL M1 = 1 mA.

TYPICAL RESULTS:

| SIGNAL (Hz) | M1 (mA) |
|---|---|
| 0 | .02 |
| 100 | .1 |
| 200 | .24 |
| 300 | .34 |
| 400 | .44 |
| 500 | .55 |
| 600 | .65 |
| 700 | .77 |
| 800 | .85 |
| 900 | .95 |
| 1000 | 1.00 |

# PULSE GENERATOR

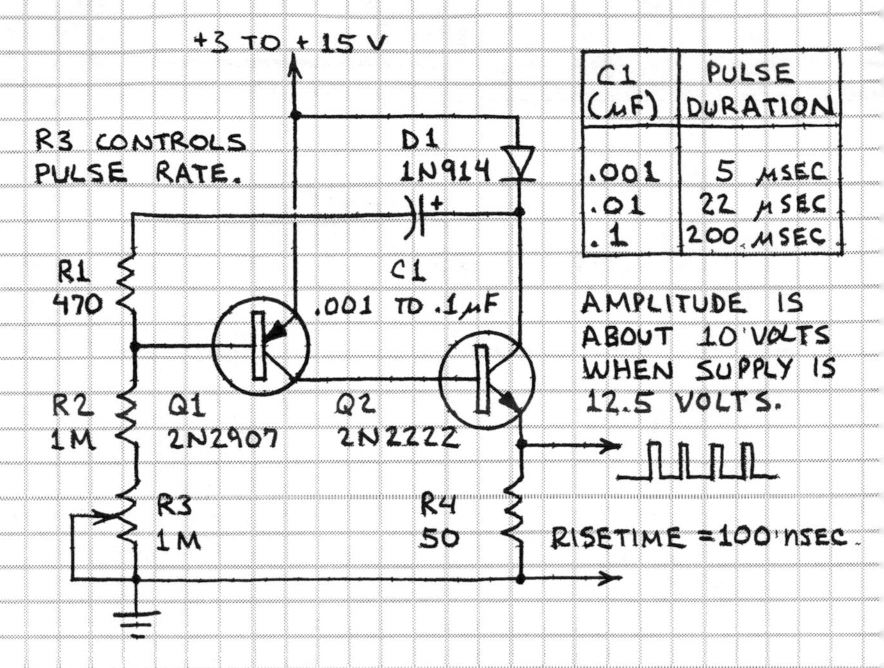


| C1 (μF) | PULSE DURATION |
| --- | --- |
| .001 | 5 μSEC |
| .01 | 22 μSEC |
| .1 | 200 μSEC |

# DC METER AMPLIFIER

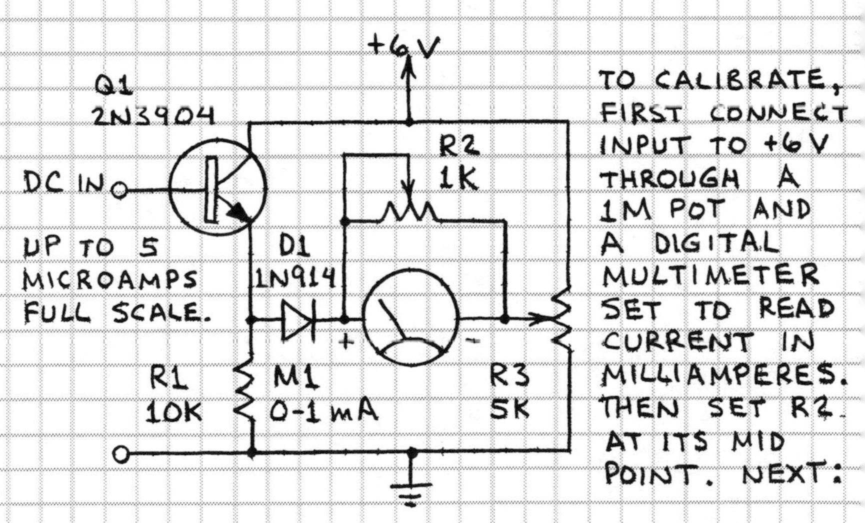


TO CALIBRATE, FIRST CONNECT INPUT TO +6V THROUGH A 1M POT AND A DIGITAL MULTIMETER SET TO READ CURRENT IN MILLIAMPERES. THEN SET R2 AT ITS MID POINT. NEXT:

1. SET 1M POT FOR DESIRED CURRENT.
2. ADJUST R3 UNTIL M1 INDICATES 1 mA.
3. REPEAT STEPS 1 AND 2.
4. ADJUST R2 UNTIL M1 INDICATES 1 mA.

# LIGHT-ACTIVATED FLASHER

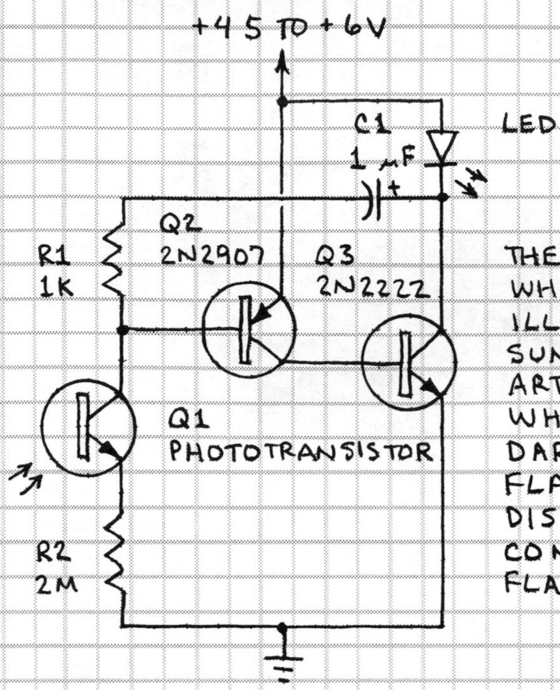


THE LED FLASHES WHEN Q1 IS ILLUMINATED BY SUNLIGHT OR ARTIFICIAL LIGHT. WHEN Q1 IS DARK, THE FLASHER IS DISABLED. C1 CONTROLS THE FLASH RATE.

# DARK-ACTIVATED FLASHER

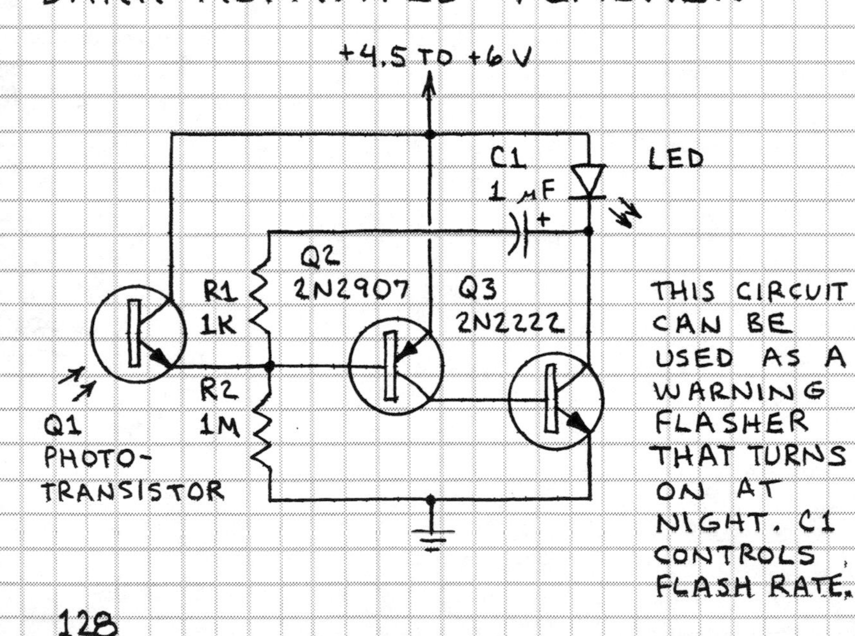


THIS CIRCUIT CAN BE USED AS A WARNING FLASHER THAT TURNS ON AT NIGHT. C1 CONTROLS FLASH RATE.

# HIGH-BRIGHTNESS FLASHER

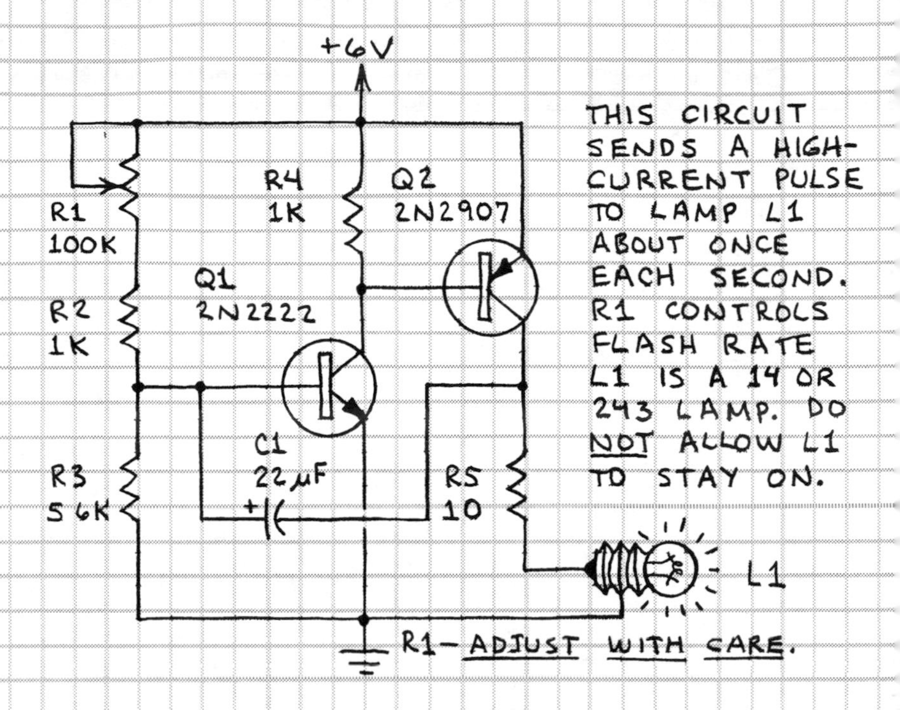


THIS CIRCUIT SENDS A HIGH-CURRENT PULSE TO LAMP L1 ABOUT ONCE EACH SECOND. R1 CONTROLS FLASH RATE. L1 IS A 14 OR 243 LAMP. DO NOT ALLOW L1 TO STAY ON.

# LED TRANSMITTER/RECEIVER

USE HIGH-OUTPUT INFRARED LED.

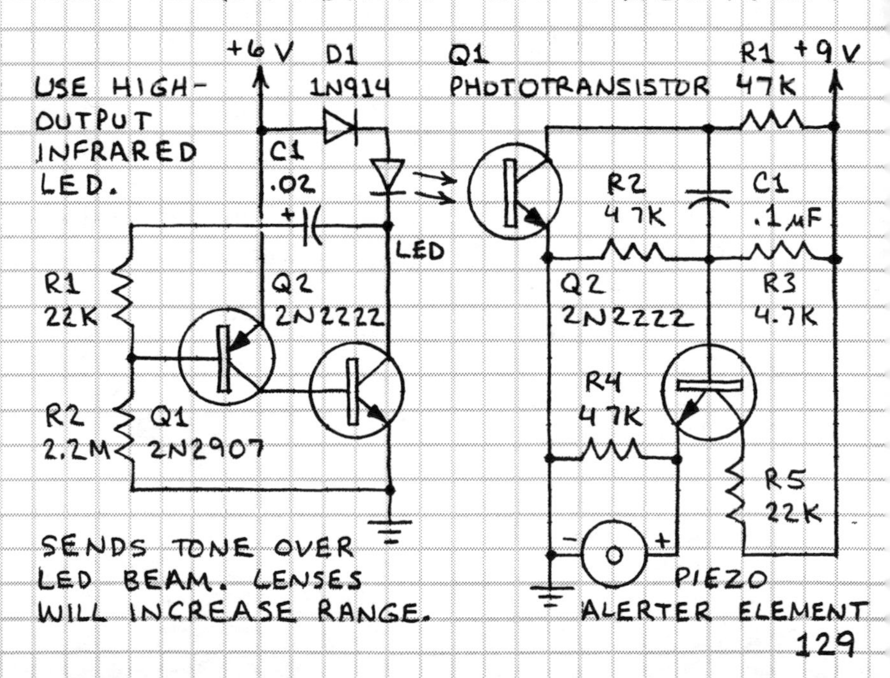


SENDS TONE OVER LED BEAM. LENSES WILL INCREASE RANGE.

# RESISTOR-TRANSISTOR LOGIC

THESE LOGIC CIRCUITS CAN BE USED TO TEACH BASICS OF DIGITAL LOGIC AND IN PRACTICAL APPLICATIONS.

## OR GATE

0 = GROUND
1 = +6V

| A | B | LED |
|---|---|---|
| 0 | 0 | OFF |
| 0 | 1 | ON |
| 1 | 0 | ON |
| 1 | 1 | ON |

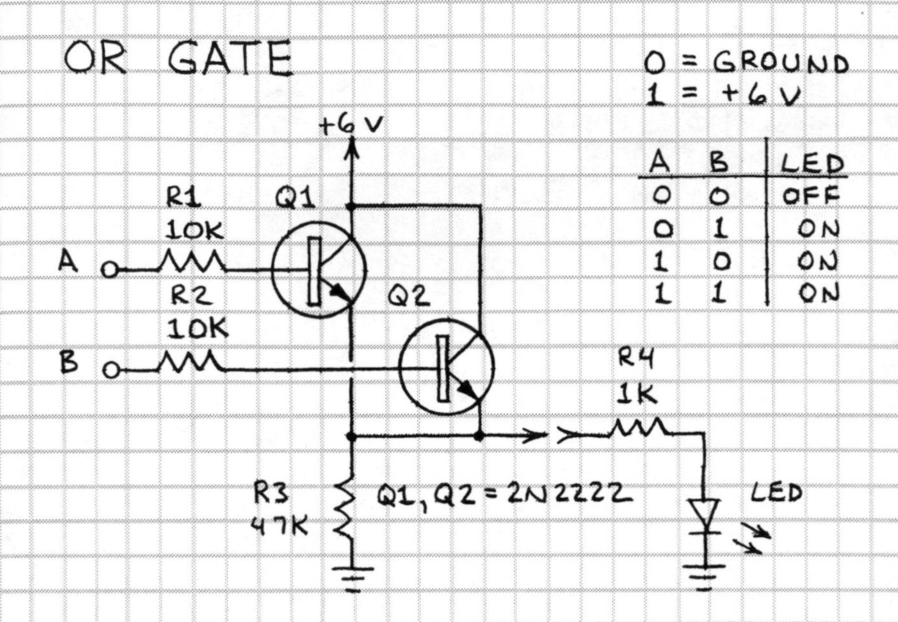


## NOR GATE

| A | B | LED |
|---|---|---|
| 0 | 0 | ON |
| 0 | 1 | OFF |
| 1 | 0 | OFF |
| 1 | 1 | OFF |

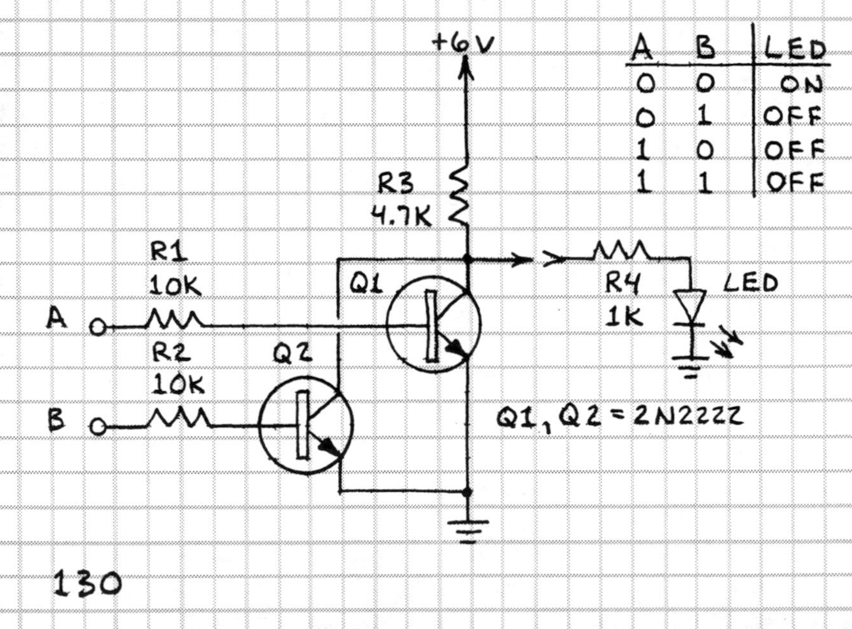

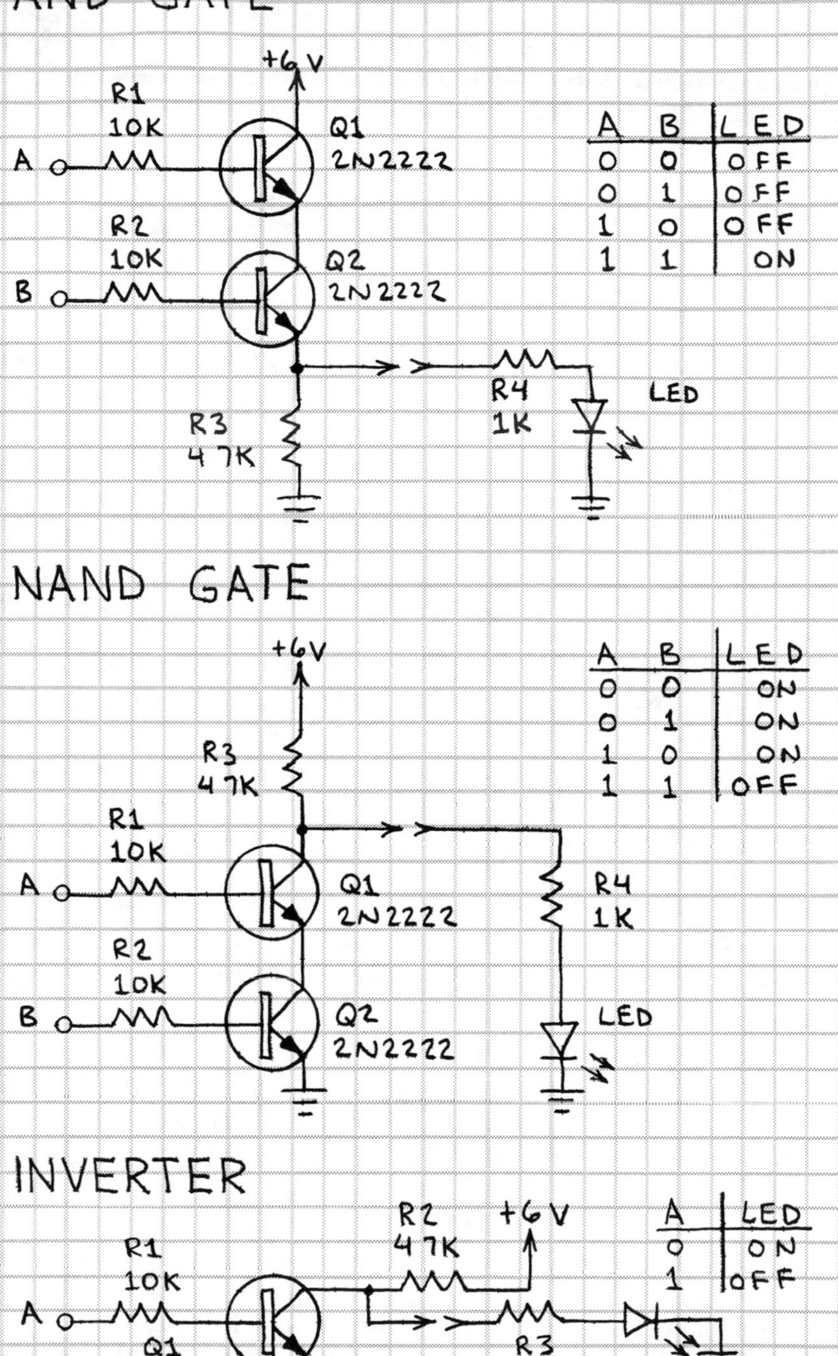
AND GATE
+6V
R1
10K
A
Q1
2N2222
R2
10K
B
Q2
2N2222
R3
47K
R4
1K
LED
A B LED
0 0 OFF
0 1 OFF
1 0 OFF
1 1 ON
NAND GATE
+6V
R3
47K
R1
10K
A
Q1
2N2222
R2
10K
B
Q2
2N2222
R4
1K
LED
A B LED
0 0 ON
0 1 ON
1 0 ON
1 1 OFF
INVERTER
R1
10K
A
Q1
2N2222
R2
47K
+6V
R3
1K
LED
A LED
0 ON
1 OFF

# JUNCTION FETS

A JUNCTION FIELD-EFFECT TRANSISTOR (FET) IS A 3-TERMINAL SEMICONDUCTOR DEVICE IN WHICH A SMALL VOLTAGE AT ONE TERMINAL CAN CONTROL A CURRENT FLOWING BETWEEN THE SECOND AND THIRD TERMINAL. FETS CAN FUNCTION AS BOTH AMPLIFIERS AND SWITCHES. THE PRINCIPLE ADVANTAGE OF THE FET IS ITS VERY HIGH INPUT (GATE) IMPEDANCE. FETS ARE CLASSIFIED AS EITHER N- OR P-CHANNEL ACCORDING TO THE DOPING OF THE CURRENT-CARRYING CHANNEL REGION.

# BASIC FET SWITCHES (N-FET)

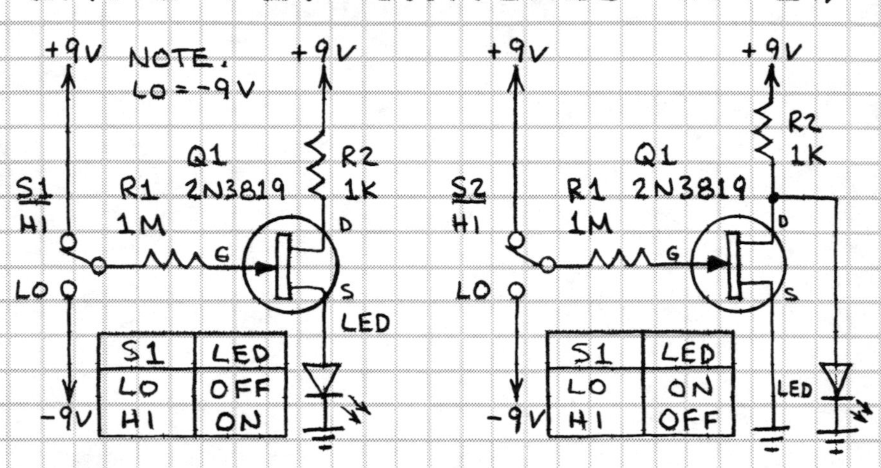


# BASIC FET AMPLIFIER (N-FET)

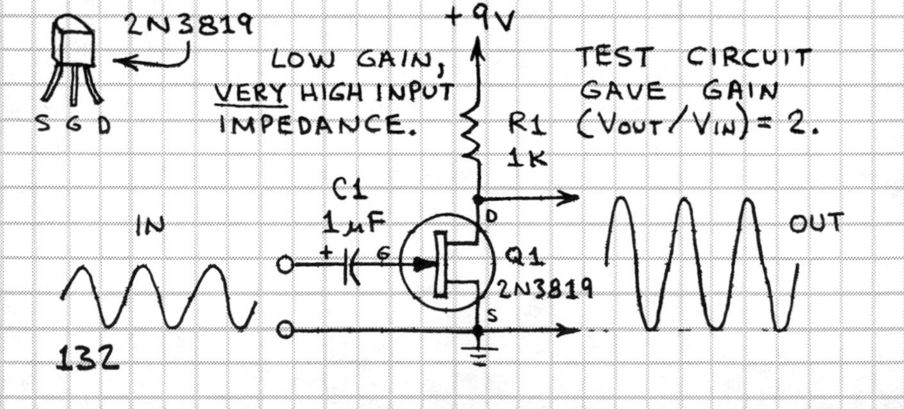

# HI-Z MICROPHONE PREAMPLIFIER

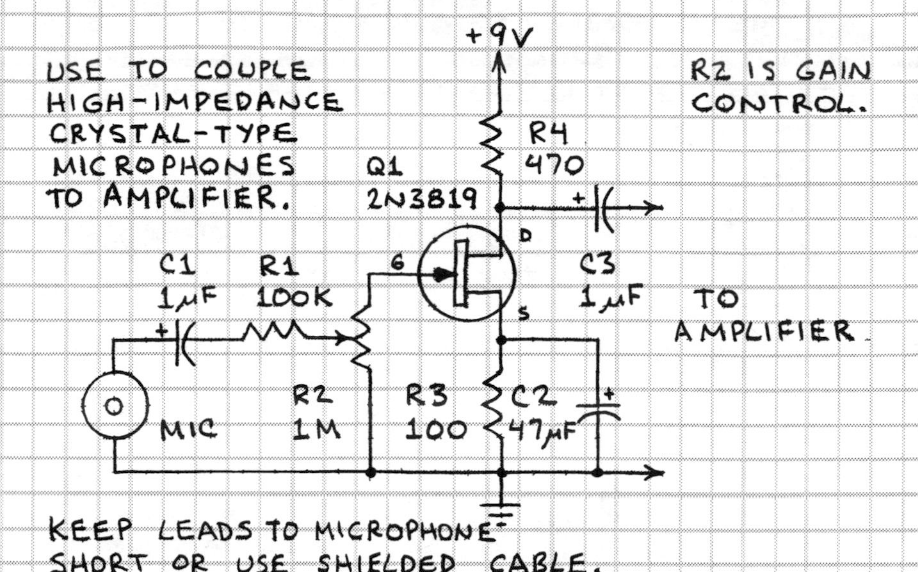


KEEP LEADS TO MICROPHONE SHORT OR USE SHIELDED CABLE.

# HI-Z AUDIO MIXER

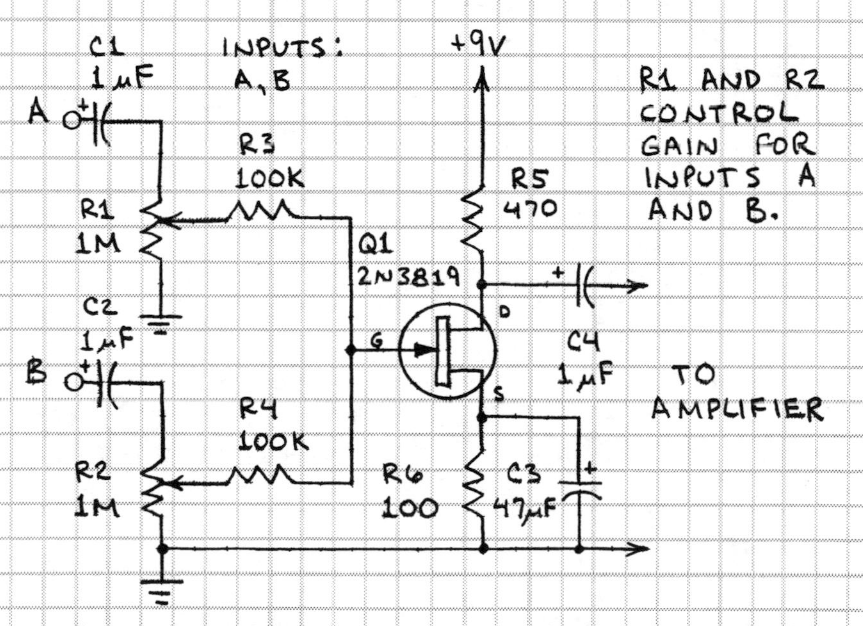


USE TO COMBINE SIGNALS FROM TWO OR MORE MICROPHONES, PREAMPLIFIERS, ETC.

# POWER MOSFETS

A METAL-OXIDE-SEMICONDUCTOR FET (MOSFET) HAS A GATE WHICH IS INSULATED FROM THE CHANNEL BY A VERY THIN GLASSY OXIDE. THEREFORE THE INPUT IMPEDANCE OF THE MOSFET IS CONSIDERABLY HIGHER THAN THAT OF THE STANDARD FET. POWER MOSFETS HAVE A VERY LOW RESISTANCE CHANNEL. THEREFORE THEY CAN CONTROL MUCH MORE CURRENT THAN FETS.

## ON-AFTER-DELAY TIMER

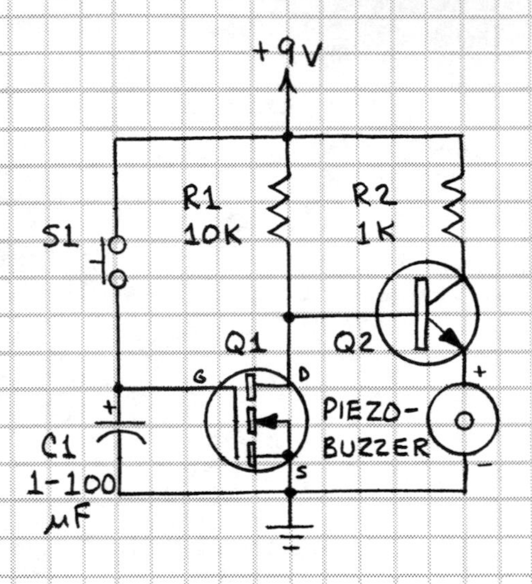


PRESS S1 TO CHARGE C1. THE PIEZO-BUZZER EMITS TONE <u>AFTER</u> C1 SELF DISCHARGES. LARGE VALUES FOR C1 INCREASE THE DELAY. PLACE LARGE VALUE RESISTOR ACROSS C1 TO REDUCE DELAY
Q1 - POWER MOSFET.
Q2 - 2N2222.

## ON-DURING-DELAY TIMER

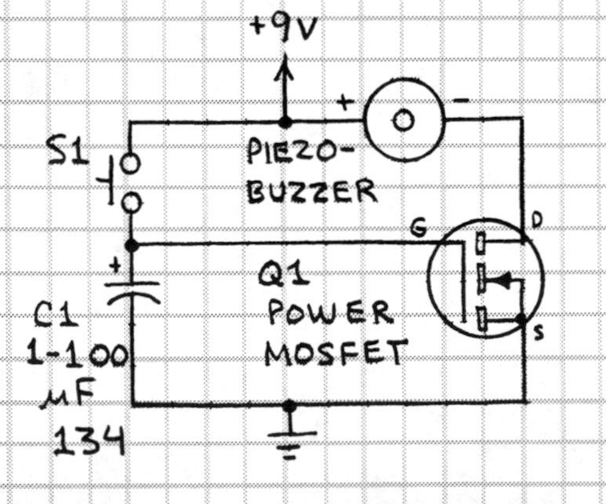


PRESS S1 TO CHARGE C1. THE PIEZO-BUZZER EMITS TONE <u>UNTIL</u> C1 SELF DISCHARGES. INCREASE C1 TO INCREASE DELAY. RESISTOR ACROSS C1 WILL REDUCE DELAY.

# HI-Z SPEAKER AMPLIFIER

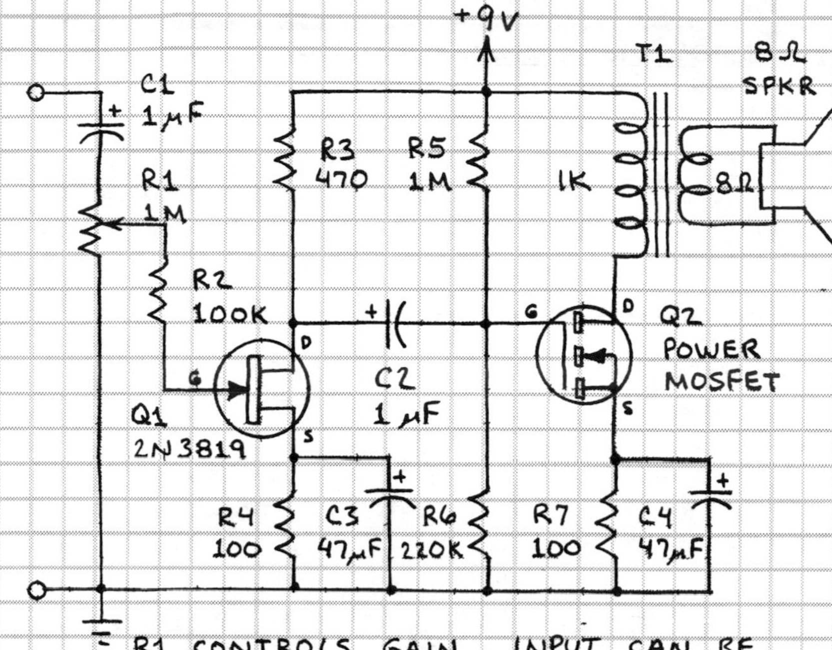


R1 CONTROLS GAIN. INPUT CAN BE HIGH-IMPEDANCE MICROPHONE, RADIO, ETC.

# DUAL LED FLASHER

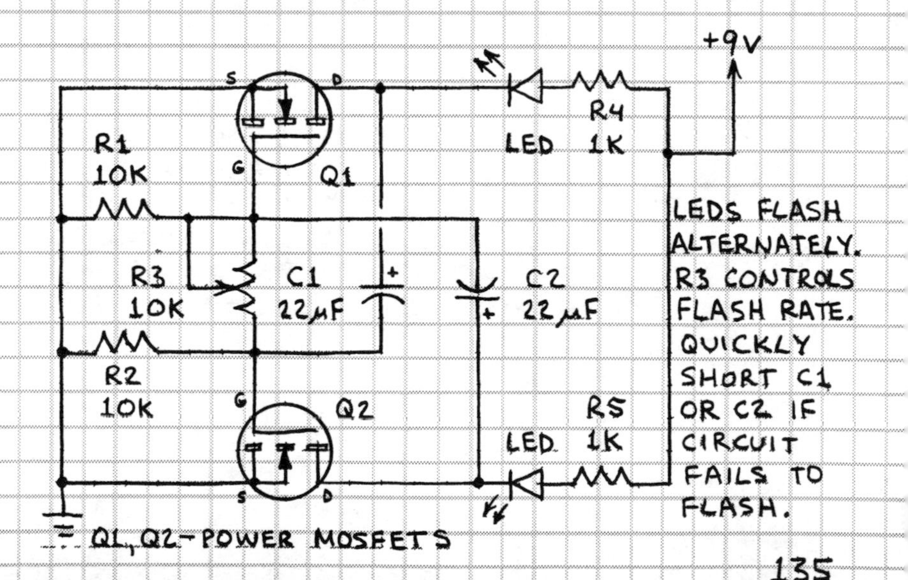


LEDS FLASH ALTERNATELY. R3 CONTROLS FLASH RATE. QUICKLY SHORT C1 OR C2 IF CIRCUIT FAILS TO FLASH.

# UNIJUNCTION TRANSISTORS

THE UNIJUNCTION TRANSISTOR (UJT) IS A VOLTAGE-CONTROLLED SWITCH AND NOT A TRUE TRANSISTOR. THE UJT IS WELL SUITED FOR MANY OSCILLATOR APPLICATIONS.

## BASIC UJT OSCILLATOR

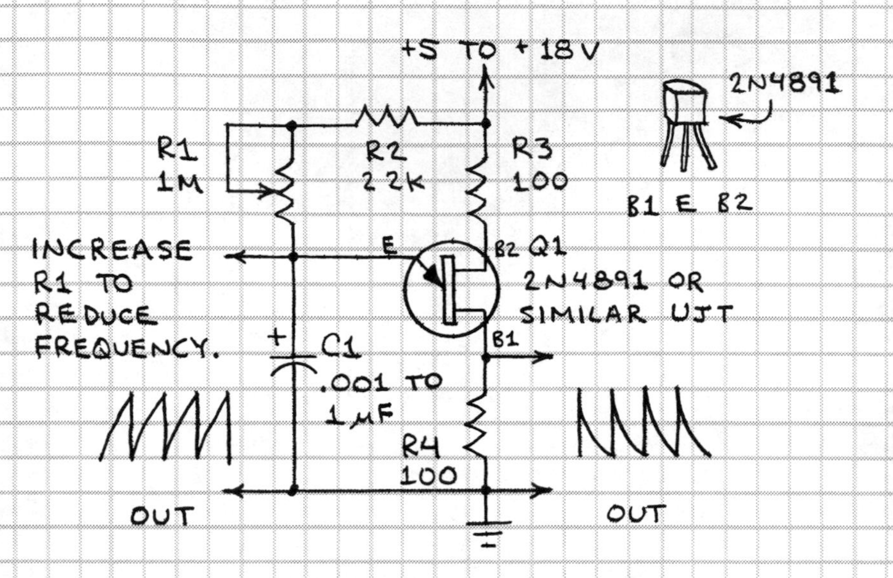


## LOW-VOLTAGE INDICATOR

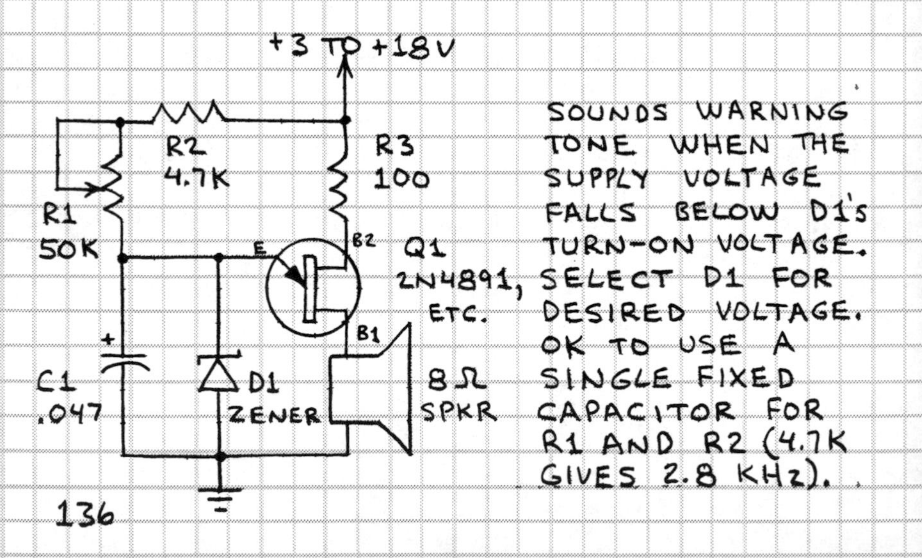


SOUNDS WARNING TONE WHEN THE SUPPLY VOLTAGE FALLS BELOW D1'S TURN-ON VOLTAGE. SELECT D1 FOR DESIRED VOLTAGE. OK TO USE A SINGLE FIXED CAPACITOR FOR R1 AND R2 (4.7K GIVES 2.8 KHz).

# SOUND-EFFECTS GENERATOR

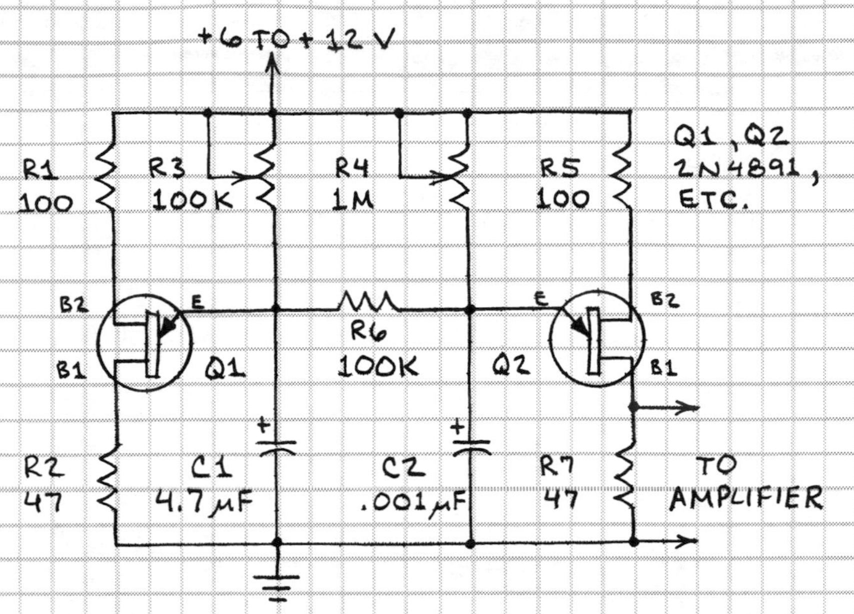


THIS CIRCUIT GENERATES CHIRPS HAVING A FREQUENCY CONTROLLED BY R4. R3 CONTROLS RATE.

# 1-MINUTE TIMER

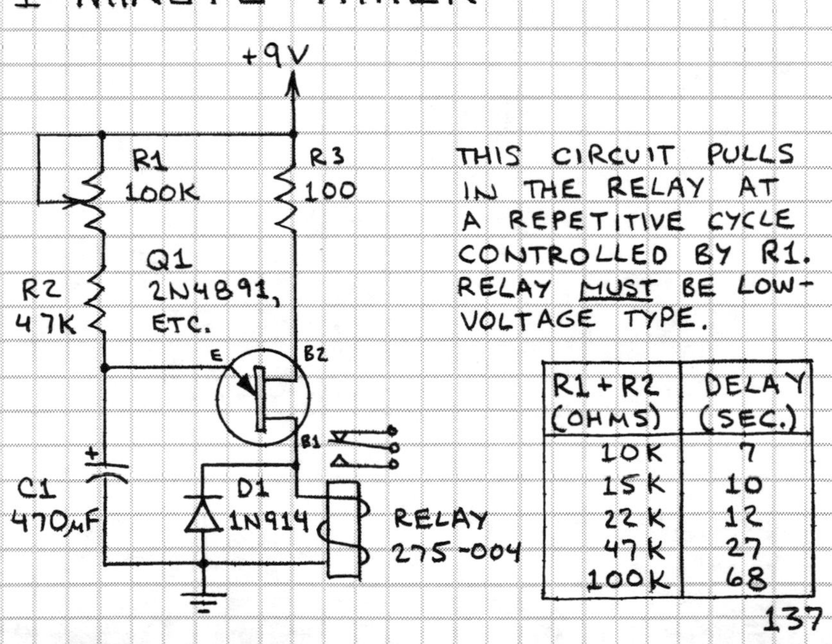


THIS CIRCUIT PULLS IN THE RELAY AT A REPETITIVE CYCLE CONTROLLED BY R1. RELAY *MUST* BE LOW-VOLTAGE TYPE.

| R1 + R2 (OHMS) | DELAY (SEC.) |
|---|---|
| 10K | 7 |
| 15K | 10 |
| 22K | 12 |
| 47K | 27 |
| 100K | 68 |

# PIEZOELECTRIC BUZZERS

PIEZO BUZZERS DELIVER EAR-PIERCING TONE AT LOW DRIVE CURRENT AND VOLTAGE.

**CAUTION:** USE EAR PROTECTORS WHEN EXPERIMENTING WITH PIEZO BUZZERS AT CLOSE RANGE FOR MORE THAN BRIEF INTERVALS.

## BELL

## VOLUME CONTROL

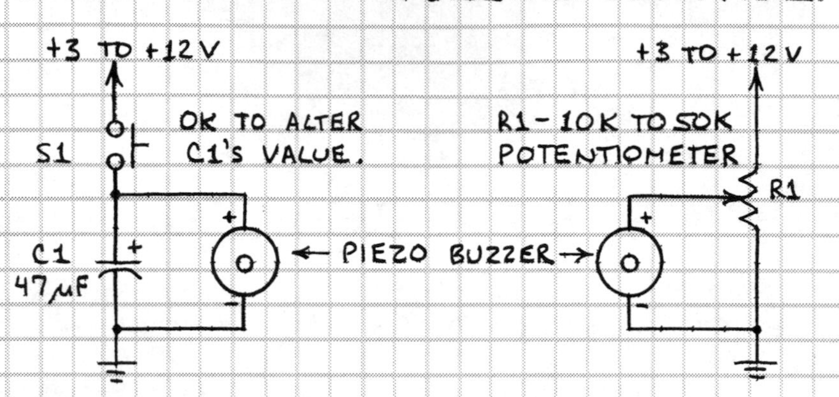


PRESS AND RELEASE S1 TO SIMULATE BELL.

R1 CONTROLS VOLUME.

## LOGIC INTERFACES

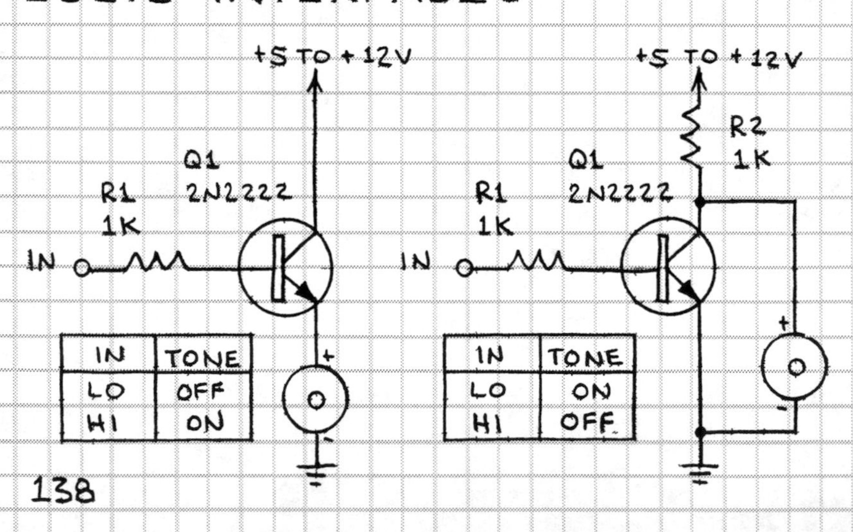


| IN | TONE |
|---|---|
| LO | OFF |
| HI | ON |

| IN | TONE |
|---|---|
| LO | ON |
| HI | OFF |

# PIEZO-ELEMENT DRIVERS

## FIXED TONE

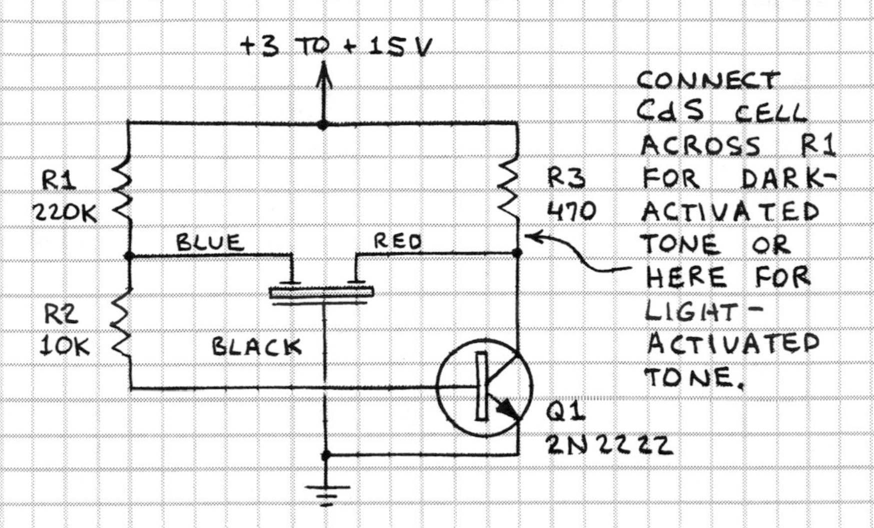


## ADJUSTABLE FREQUENCY

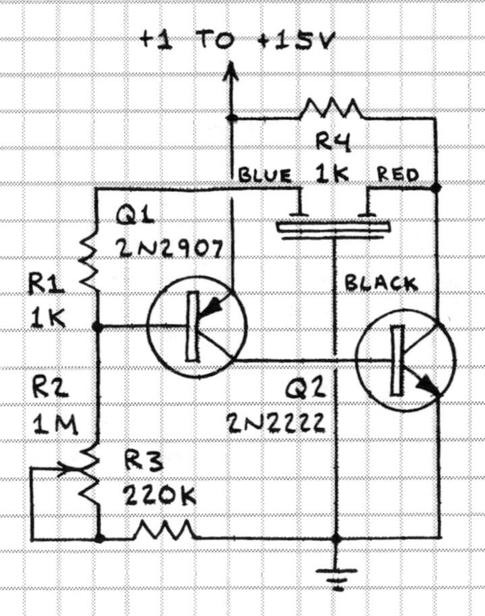


THIS CIRCUIT CAN BE EASILY MINIATURIZED. R2 CONTROLS FREQUENCY.

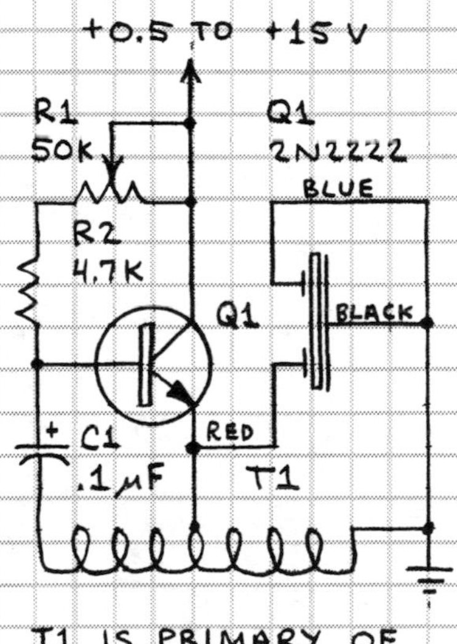


T1 IS PRIMARY OF CENTER-TAPPED AUDIO TRANSFORMER (RADIO SHACK 273-1380). R1 CONTROLS FREQUENCY.

# SILICON-CONTROLLED RECTIFIERS

THE SILICON-CONTROLLED RECTIFIER (SCR) IS A TRUE SOLID-STATE ON-OFF SWITCH. THE SCR IS SWITCHED ON BY A SMALL CURRENT AT ITS GATE TERMINAL. THE SCR WILL REMAIN ON UNTIL THE CURRENT FLOWING THROUGH IT FALLS BELOW A MINIMUM LEVEL ($I_H$ OR HOLDING CURRENT).

# LATCHING PUSHBUTTON SWITCH

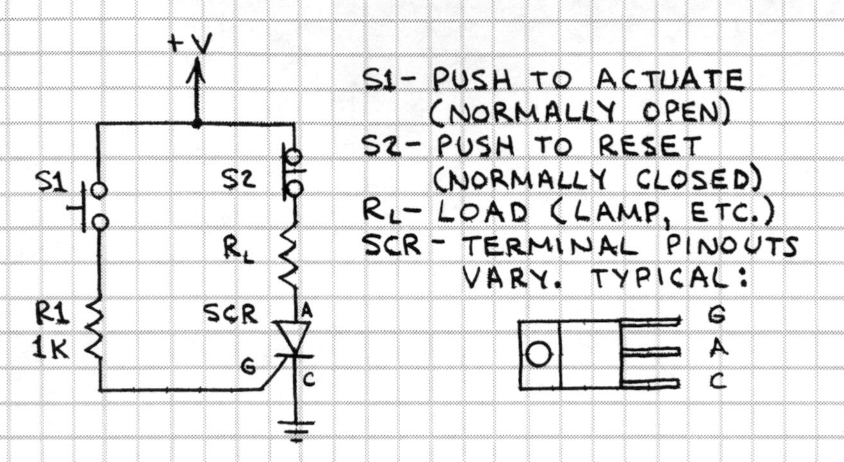


# LIGHT-ACTIVATED RELAY

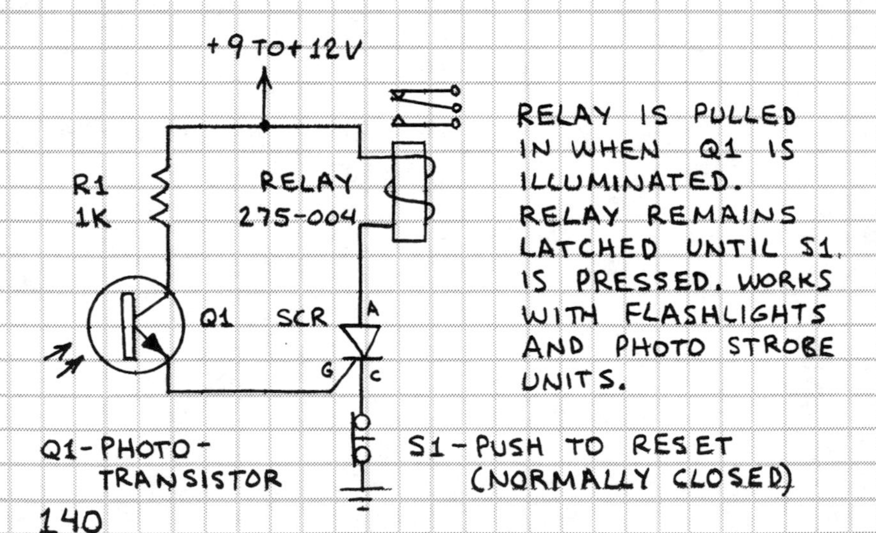

# RELAXATION OSCILLATOR

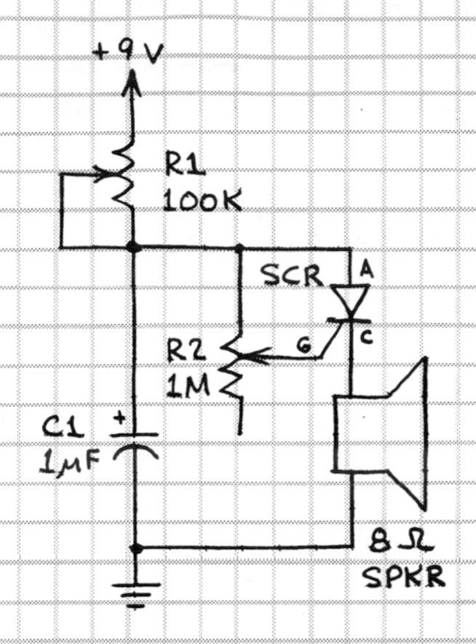


C1 IS CHARGED THROUGH R1 UNTIL ITS CHARGE IS HIGH ENOUGH TO SWITCH ON THE SCR THROUGH R2. C1 THEN DISCHARGES THROUGH THE SCR AND THE SPEAKER. R1 CONTROLS THE REPETION RATE.

NOTE: SOME SCRs REQUIRE CAREFUL ADJUSTMENT OF R2.

# DC MOTOR SPEED CONTROLLER

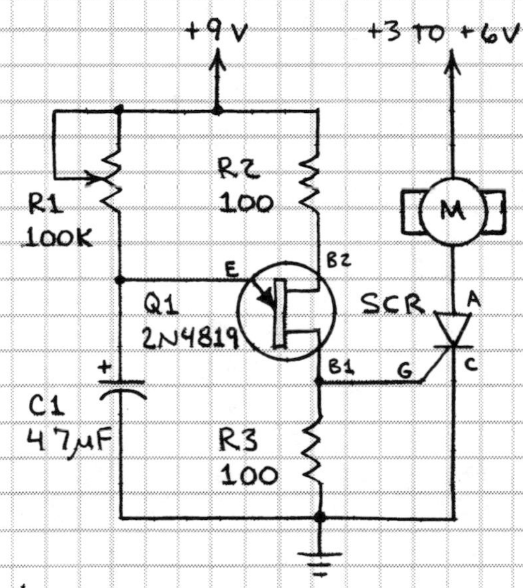


THIS CIRCUIT WILL VARY THE SPEED OF SELECTED* DC MOTORS. R4 CONTROLS THE SPEED. AT SLOW PULSE RATES FROM THE UJT OSCILLATOR, THE MOTOR WILL ROTATE IN BURSTS. FOR BEST RESULTS, USE A SEPARATE POWER SUPPLY FOR THE MOTOR.

*CHECK MOTOR WITH THIS CIRCUIT. IF LED FLASHES ON AND OFF WHEN SHAFT OF MOTOR IS ROTATED, IT WILL PROBABLY WORK.

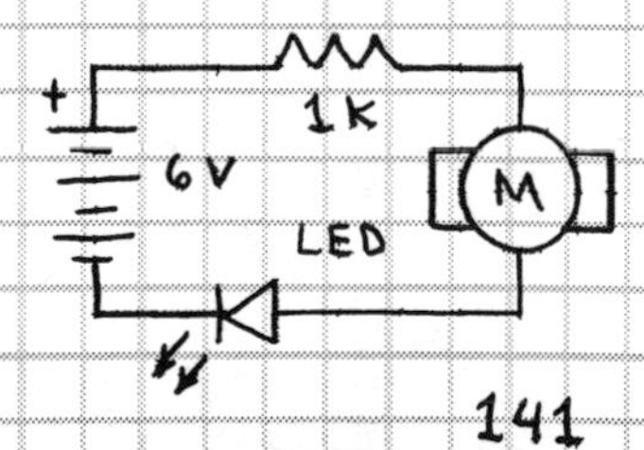

# TRIACS

THE TRIAC IS A SOLID-STATE ON-OFF SWITCH THAT CAN CONTROL ALTERNATING CURRENT. IT IS ELECTRONICALLY EQUAL TO TWO SCRs CONNECTED IN REVERSE-PARALLEL.

<u>WARNING</u>: TRIACS ARE DESIGNED FOR AC OPERATION. USE COMMON SENSE SAFETY PRECAUTIONS WHEN WORKING WITH CIRCUITS THAT USE HOUSEHOLD LINE CURRENT. <u>ALL</u> CONNECTIONS <u>MUST</u> BE WELL INSULATED. <u>NEVER</u> WORK ON AN AC LINE POWERED CIRCUIT WHEN THE POWER CORD PLUG IS INSERTED IN A WALL SOCKET.

## TRIAC SWITCH BUFFER

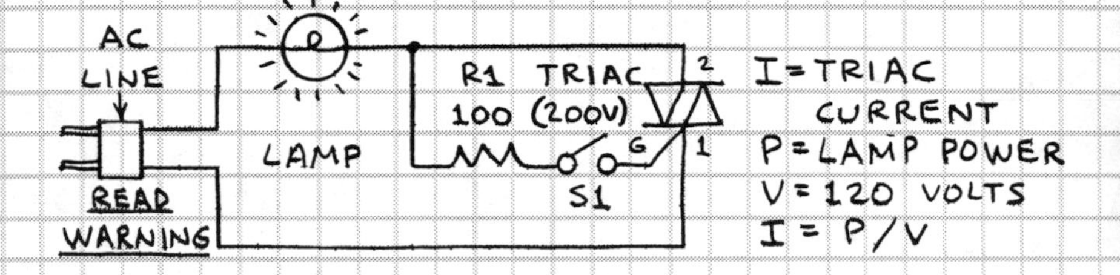


## LAMP DIMMER

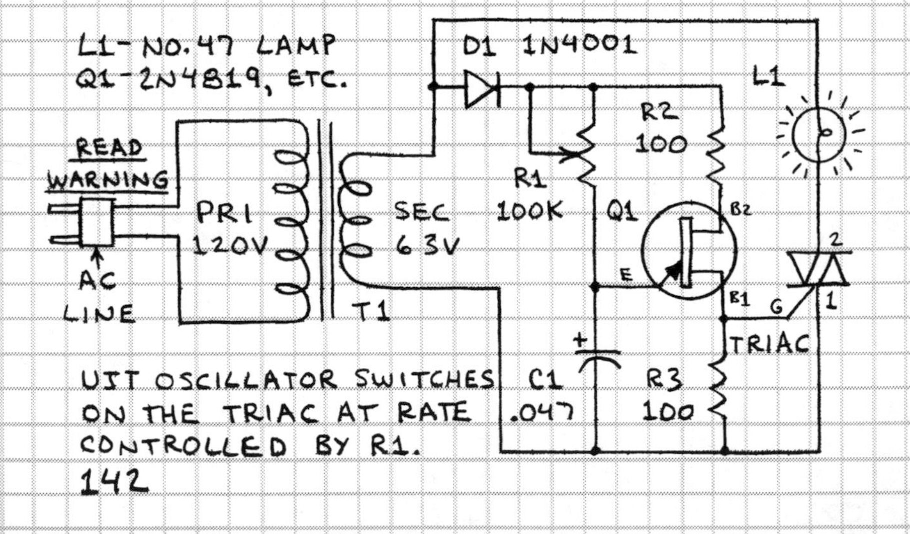


UJT OSCILLATOR SWITCHES ON THE TRIAC AT RATE CONTROLLED BY R1.

# IV. DIGITAL LOGIC CIRCUITS

## OVERVIEW

DIGITAL ELECTRONICS IS THE TECHNOLOGY THAT MAKES POSSIBLE ELECTRONIC WATCHES, CLOCKS, CALCULATORS, COMPUTERS AND MANY OTHER DEVICES. THE CIRCUITS IN THIS SECTION PROVIDE A BASIC INTRODUCTION TO DIGITAL LOGIC CIRCUITS. MANY OF THE CIRCUITS ARE SELF-FUNCTIONING AND REQUIRE NO ADDITIONAL COMPONENTS OR CIRCUITS. SOME CIRCUITS ARE DESIGNED TO WORK WITH OTHER LOGIC CIRCUITS. TO SIMPLIFY THIS PROCEDURE AND TO ENCOURAGE EXPERIMENTATION AND DO-IT-YOURSELF CIRCUIT DESIGN, THIS SECTION INCLUDES MANY METHODS FOR INTERFACING LOGIC CIRCUITS WITH ONE ANOTHER AND WITH EXTERNAL COMPONENTS. AS FOR THE CIRCUITS, EQUAL ATTENTION IS GIVEN TO TTL AND CMOS LOGIC CIRCUITS. THE CIRCUITS CAN BE ADAPTED FOR USE WITH OTHER LOGIC FAMILIES. SO THAT THE MAXIMUM NUMBER OF CIRCUITS CAN BE INCLUDED, ONLY ESSENTIAL INFORMATION IS PROVIDED.

## SWITCH LOGIC

THE SIMPLEST LOGIC CIRCUITS ARE THESE:

AND GATE

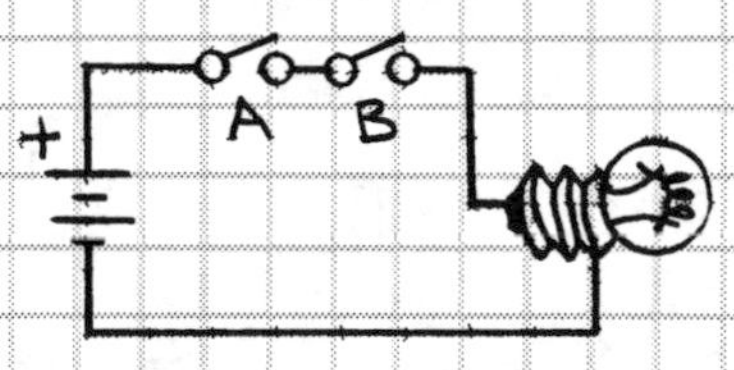

+ A B

A "AND" B = LAMP ON

OR GATE

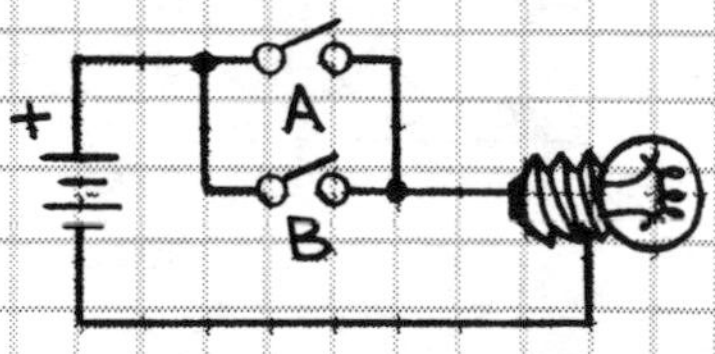

+ A B

A "OR" B = LAMP ON

# TRANSISTOR LOGIC CIRCUITS

THESE CIRCUITS SHOW HOW TRANSISTOR SWITCHES CAN BE USED TO FORM FOUR OF THE SIMPLEST LOGICAL DECISION CIRCUITS OR GATES. EACH CIRCUIT INCLUDES A TRUTH TABLE THAT GIVES THE OUTPUT FOR ALL INPUT COMBINATIONS.

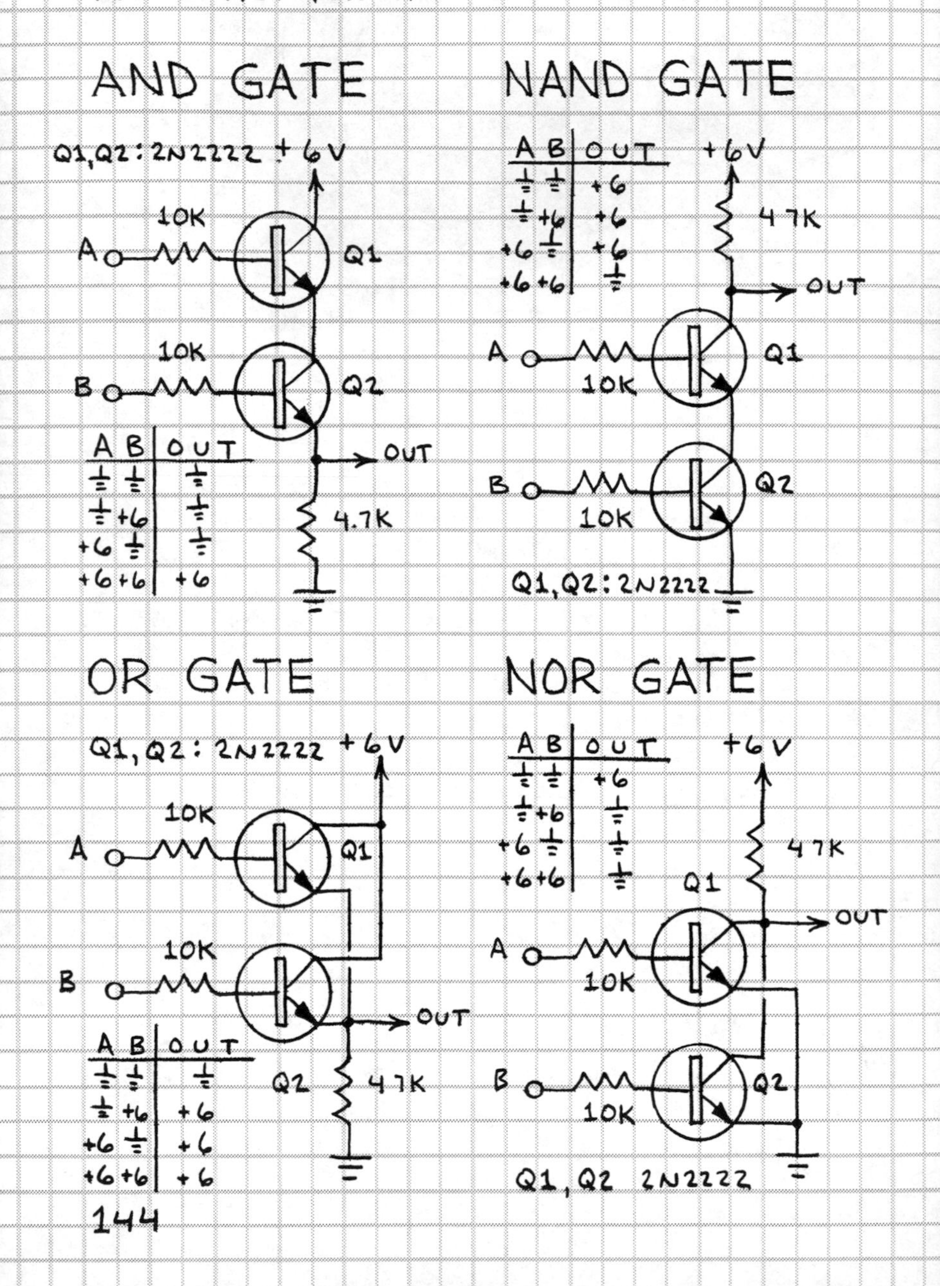


AND GATE

| A | B | OUT |
|---|---|---|
| ⏚ | ⏚ | ⏚ |
| ⏚ | +6 | ⏚ |
| +6 | ⏚ | ⏚ |
| +6 | +6 | +6 |

NAND GATE

| A | B | OUT |
|---|---|---|
| ⏚ | ⏚ | +6 |
| ⏚ | +6 | +6 |
| +6 | ⏚ | +6 |
| +6 | +6 | ⏚ |

OR GATE

| A | B | OUT |
|---|---|---|
| ⏚ | ⏚ | ⏚ |
| ⏚ | +6 | +6 |
| +6 | ⏚ | +6 |
| +6 | +6 | +6 |

NOR GATE

| A | B | OUT |
|---|---|---|
| ⏚ | ⏚ | +6 |
| ⏚ | +6 | ⏚ |
| +6 | ⏚ | ⏚ |
| +6 | +6 | ⏚ |

# BINARY (TWO-STATE) NUMBERS

THE TRUTH TABLES ON THE FACING PAGE GIVE INPUT AND OUTPUT STATES AS +6 VOLTS AND 0 VOLTS (GROUND). THESE TWO STATES CAN BE REPLACED BY THE DIGITS 1 AND 0:

| A B | AND | NAND | OR | NOR |
|---|---|---|---|---|
| 0 0 | 0 | 1 | 0 | 1 |
| 0 1 | 0 | 1 | 1 | 0 |
| 1 0 | 0 | 1 | 1 | 0 |
| 1 1 | 1 | 0 | 1 | 0 |

THE SEQUENCE OF INPUTS FORMS THE FIRST FOUR NUMBERS IN THE BINARY SYSTEM.

OTHER 2-INPUT LOGIC GATES INCLUDE:

| A B | EXCLUSIVE OR | EXCLUSIVE NOR |
|---|---|---|
| 0 0 | 0 | 1 |
| 0 1 | 1 | 0 |
| 1 0 | 1 | 0 |
| 1 1 | 0 | 1 |

A BINARY DIGIT (0 OR 1) IS CALLED A BIT. PATTERNS OF BITS CAN REPRESENT DECIMAL NUMBERS, LETTERS OF THE ALPHABET, VOLTAGES AND OTHER INFORMATION. FOR EXAMPLE:

| DECIMAL | BINARY | BCD | |
|---|---|---|---|
| 0 | 0000 | 0000 | 0000 |
| 1 | 0001 | 0000 | 0001 |
| 2 | 0010 | 0000 | 0010 |
| 3 | 0011 | 0000 | 0011 |
| 4 | 0100 | 0000 | 0100 |
| 5 | 0101 | 0000 | 0101 |
| 6 | 0110 | 0000 | 0110 |
| 7 | 0111 | 0000 | 0111 |
| 8 | 1000 | 0000 | 1000 |
| 9 | 1001 | 0000 | 1001 |
| 10 | 1010 | 0001 | 0000 |
| 11 | 1011 | 0001 | 0001 |
| 12 | 1100 | 0001 | 0010 |
| 13 | 1101 | 0001 | 0011 |
| 14 | 1110 | 0001 | 0100 |
| 15 | 1111 | 0001 | 0101 |

BCD IS BINARY-CODED DECIMAL. BCD PROVIDES A SHORTCUT WAY TO DISPLAY DECIMAL NUMBERS ON CALCULATOR AND WATCH READOUTS. EACH DECIMAL DIGIT IS REPRESENTED BY 4 BITS.

NIBBLE : 4 BITS
WORD : 8 BITS

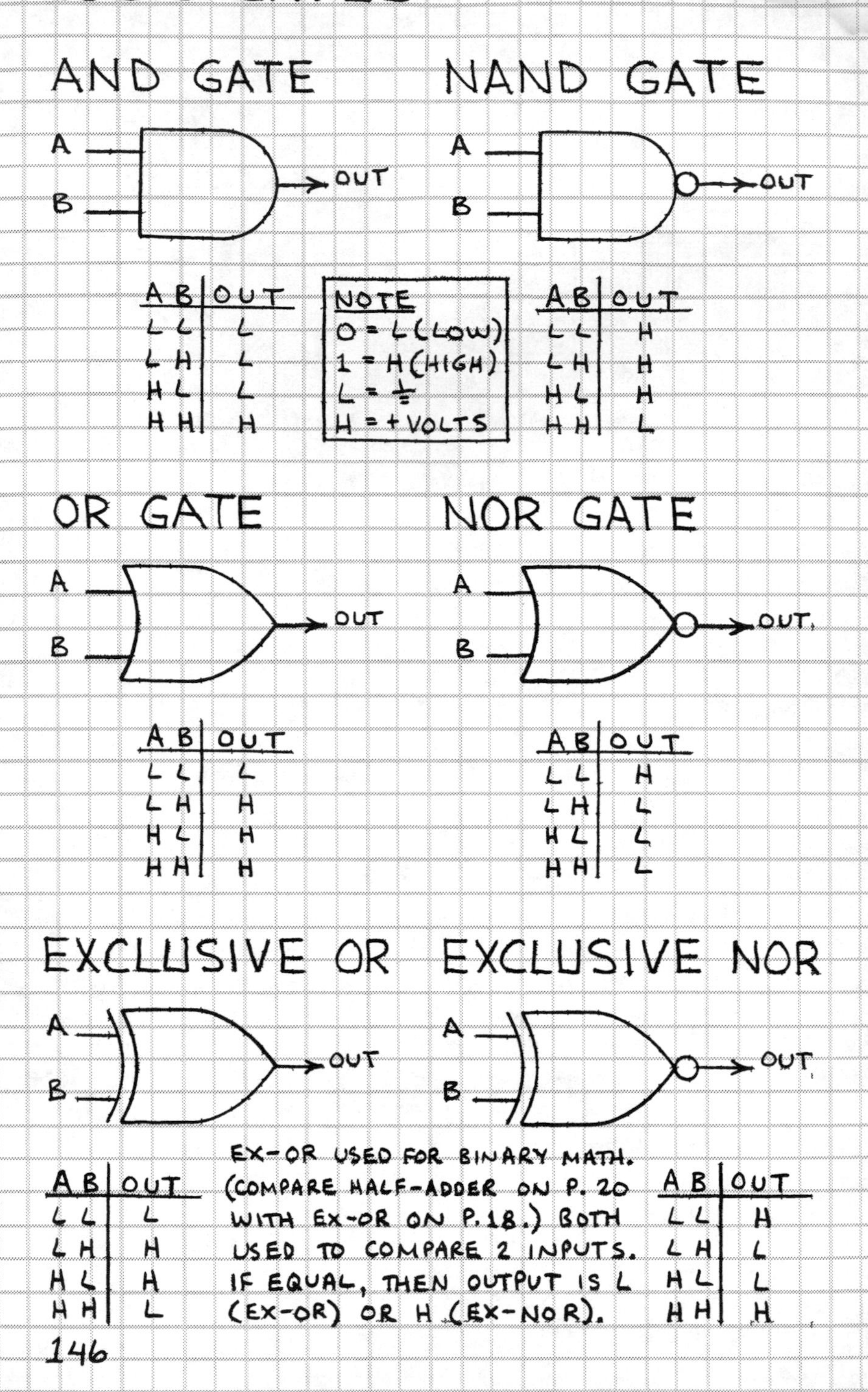
LOGIC GATES
AND GATE
A
B
OUT
A B | OUT
L L | L
L H | L
H L | L
H H | H
NAND GATE
A
B
OUT
A B | OUT
L L | H
L H | H
H L | H
H H | L
NOTE
0 = L (LOW)
1 = H (HIGH)
L = ⏚
H = + VOLTS
OR GATE
A
B
OUT
A B | OUT
L L | L
L H | H
H L | H
H H | H
NOR GATE
A
B
OUT
A B | OUT
L L | H
L H | L
H L | L
H H | L
EXCLUSIVE OR
A
B
OUT
A B | OUT
L L | L
L H | H
H L | H
H H | L
EXCLUSIVE NOR
A
B
OUT
A B | OUT
L L | H
L H | L
H L | L
H H | H
EX-OR USED FOR BINARY MATH.
(COMPARE HALF-ADDER ON P. 20
WITH EX-OR ON P. 18.) BOTH
USED TO COMPARE 2 INPUTS.
IF EQUAL, THEN OUTPUT IS L
(EX-OR) OR H (EX-NOR).

## 3-INPUT NAND 3-INPUT NOR

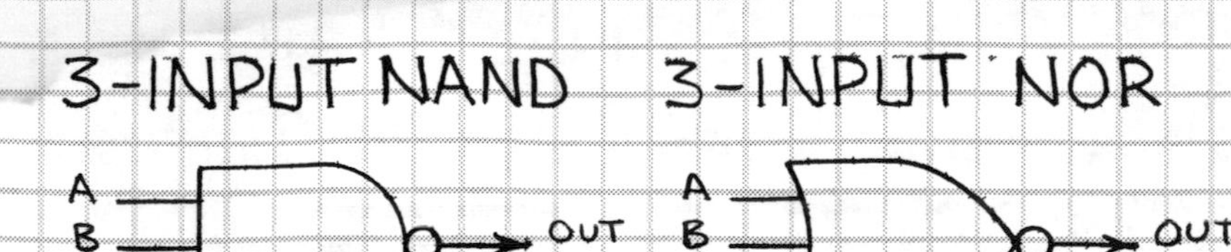


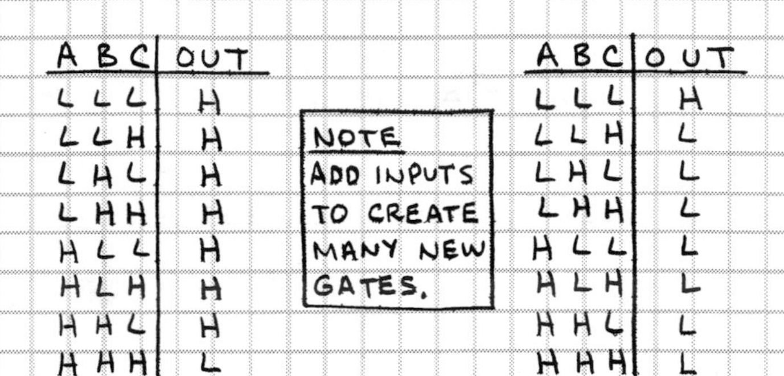

| A | B | C | OUT |
|---|---|---|---|
| L | L | L | H |
| L | L | H | H |
| L | H | L | H |
| L | H | H | H |
| H | L | L | H |
| H | L | H | H |
| H | H | L | H |
| H | H | H | L |

NOTE
ADD INPUTS TO CREATE MANY NEW GATES.

| A | B | C | OUT |
|---|---|---|---|
| L | L | L | H |
| L | L | H | L |
| L | H | L | L |
| L | H | H | L |
| H | L | L | L |
| H | L | H | L |
| H | H | L | L |
| H | H | H | L |

## BUFFER INVERTER

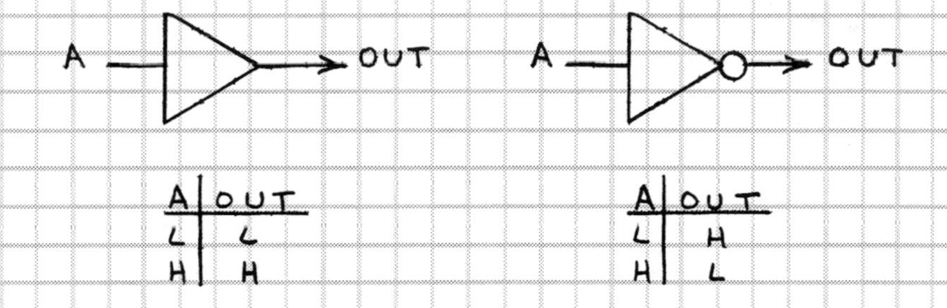


| A | OUT |
|---|---|
| L | L |
| H | H |

| A | OUT |
|---|---|
| L | H |
| H | L |

## 3-STATE LOGIC

## BUFFER INVERTER

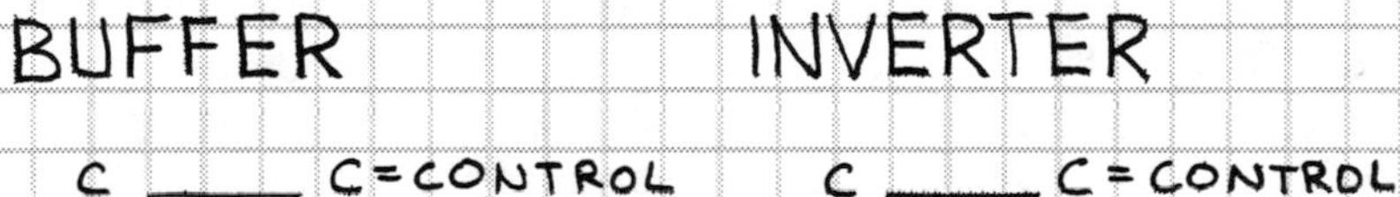


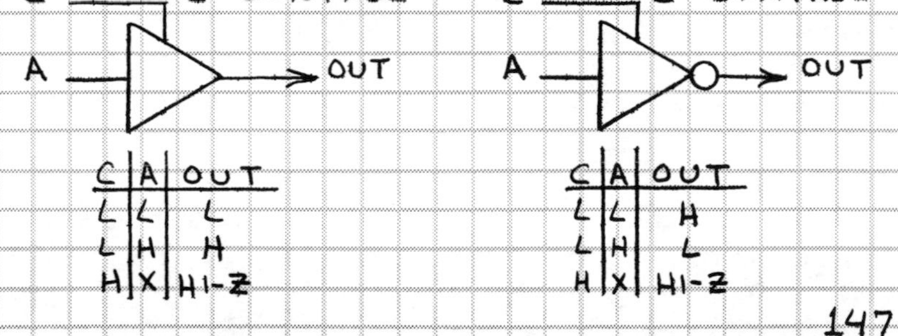


| C | A | OUT |
|---|---|---|
| L | L | L |
| L | H | H |
| H | X | HI-Z |

| C | A | OUT |
|---|---|---|
| L | L | H |
| L | H | L |
| H | X | HI-Z |

# TTL AND TTL/LS LOGIC FAMILIES

TTL (TRANSISTOR-TRANSISTOR LOGIC) AND TTL/LS (LOW-POWER SCHOTTKY) CHIPS ARE EASY TO USE AND REQUIRE NO SPECIAL HANDLING PRECAUTIONS TTL CAN CHANGE STATES 20,000,000 TIMES PER SECOND. TTL USES LOTS OF POWER, AND INDIVIDUAL GATES CONSUME 3 OR MORE MILLIAMPERES. TTL/LS IS SLIGHTLY FASTER AND USES 80% LESS POWER.

# OPERATING REQUIREMENTS

1. $V_{cc}$ (POSITIVE SUPPLY) MUST NOT EXCEED 5.25 VOLTS.

2. INPUT SIGNAL MUST NEVER EXCEED $V_{cc}$ NOR FALL BELOW GROUND.

3. UNUSED INPUTS NORMALLY ASSUME THE HIGH (H) STATE, BUT THEY MAY PICK UP STRAY SIGNALS. CONNECT THEM TO $V_{cc}$.

4. FORCE OUTPUTS OF UNUSED GATES H TO SAVE CURRENT. SEE TRUTH TABLES ON PP. 8-9.

5. TTL GATES CAUSE NOISE SPIKES ON THEIR POWER SUPPLY LEADS WHEN THEY CHANGE STATES. THESE SPIKES CAN BE REMOVED BY CONNECTING A 0.01 TO 0.1 μF DECOUPLING CAPACITOR ACROSS THE SUPPLY PINS OF TTL AND TTL/LS CHIPS. USE AT LEAST ONE CAPACITOR FOR EVERY 5 TO 10 GATE PACKAGES OR 2 TO 5 COUNTER AND REGISTER CHIPS. DECOUPLING CAPACITORS MUST HAVE SHORT LEADS AND BE CONNECTED FROM $V_{cc}$ TO GROUND AS CLOSE AS POSSIBLE TO THE DECOUPLED CHIPS.

6. AVOID LONG WIRES IN TTL AND TTL/LS CIRCUITS.

7. IF THE POWER SUPPLY IS NOT ON THE CIRCUIT BOARD, CONNECT A 1 TO 10 μF CAPACITOR ACROSS THE POWER LEADS WHERE THEY ENTER THE BOARD.

# POWER SUPPLIES

TTL CIRCUITS REQUIRE A 4.75 TO 5.25-VOLT SUPPLY. BATTERIES CAN BE USED TO POWER A FEW CHIPS. OTHERWISE A LINE-POWERED SUPPLY IS MORE ECONOMICAL AND RELIABLE.

# BATTERY POWER SUPPLIES

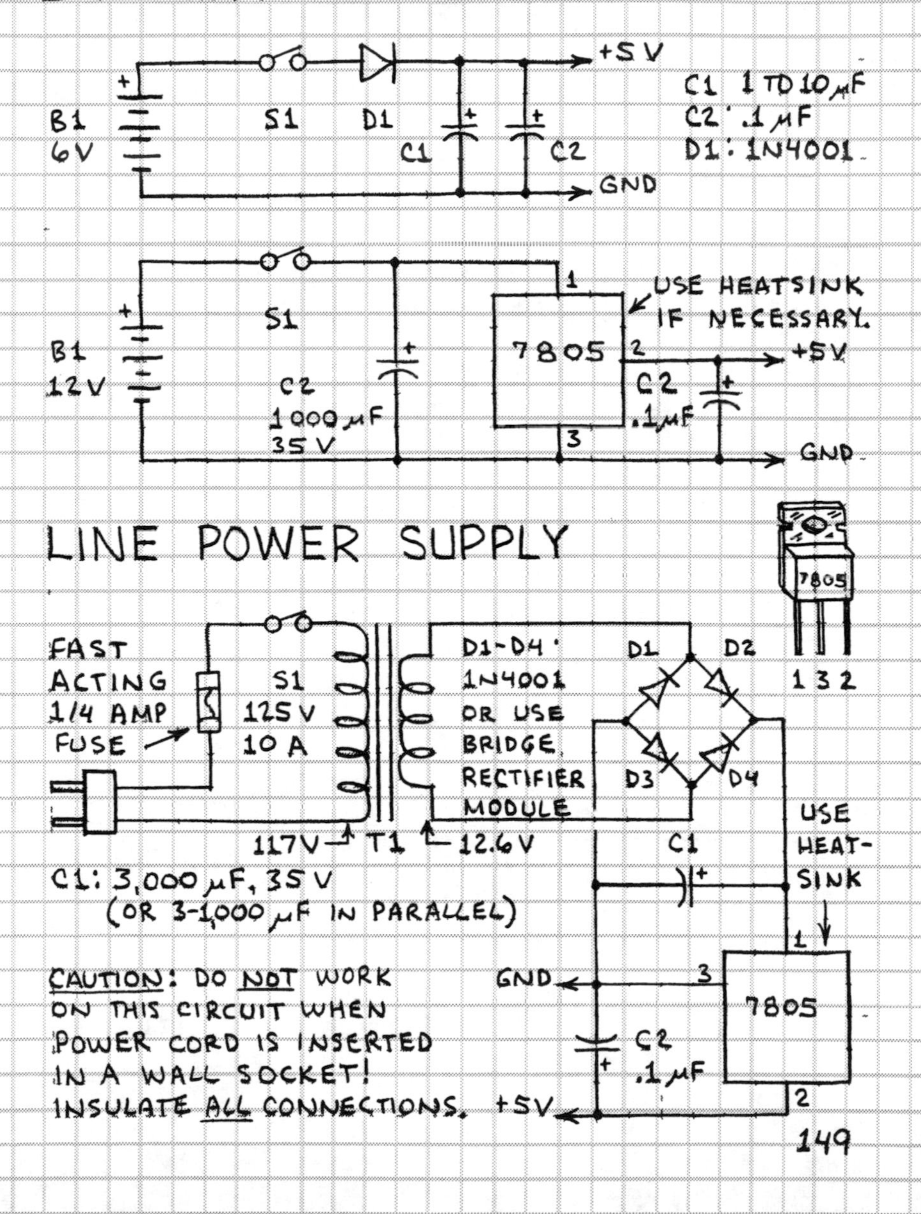


CAUTION: DO NOT WORK ON THIS CIRCUIT WHEN POWER CORD IS INSERTED IN A WALL SOCKET! INSULATE ALL CONNECTIONS.

# TTL INPUT INTERFACING

NON-TTL-TTL/LS CHIPS AND COMPONENTS CAN SUPPLY INPUT SIGNALS TO TTL-TTL/LS CHIPS IF THE OPERATING REQUIREMENTS ON PAGE 10 ARE OBSERVED. THE CIRCUITS BELOW SUPPLY CLEAN, NOISE-FREE PULSES TO TTL-TTL/LS CHIPS. THE INVERTER IN EACH CIRCUIT REPRESENTS A TTL OR TTL/LS INPUT.

## CLOCK PULSE GENERATOR

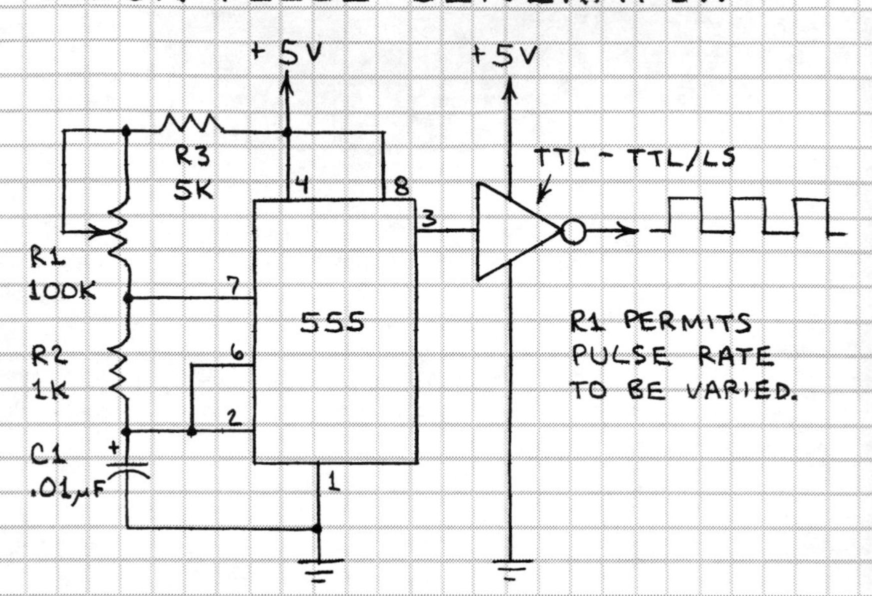


## BOUNCELESS SWITCH

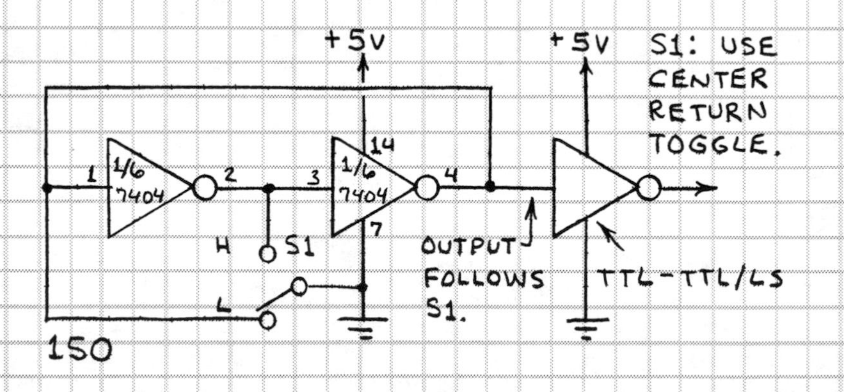

# PHOTOTRANSISTOR TO TTL

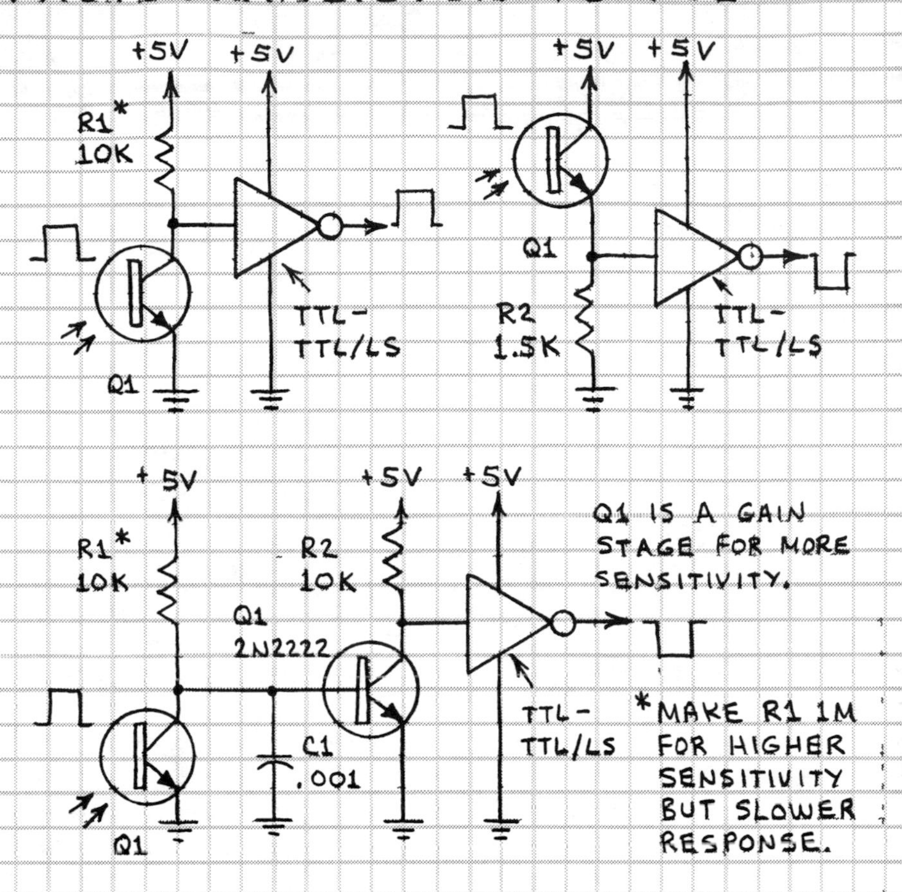


# COMPARATOR/OP-AMP TO TTL

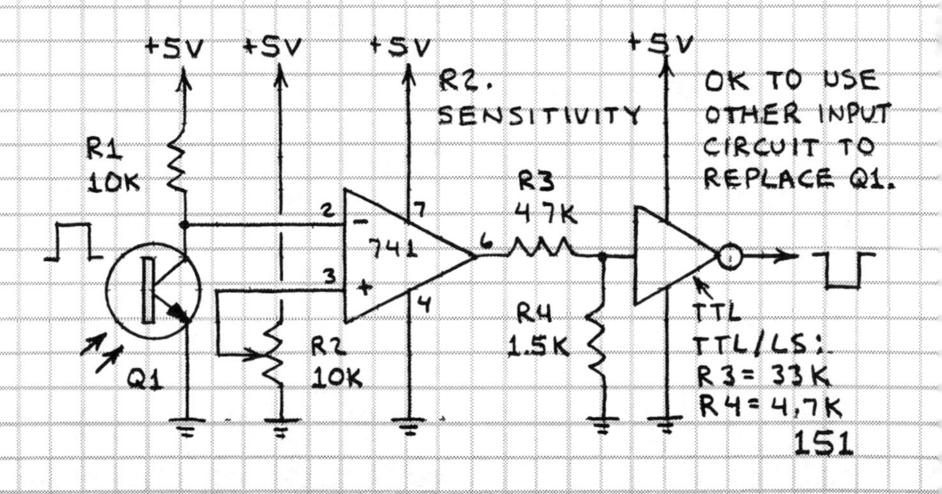

# TTL OUTPUT INTERFACING

TTL CHIPS HAVE AN OUTPUT DRIVE CURRENT OF UP TO 30 MILLIAMPERES IN A SINK (OUTPUT LOW) CONFIGURATION. SEE DATA FOR SPECIFIC CHIPS.

# LED DRIVERS

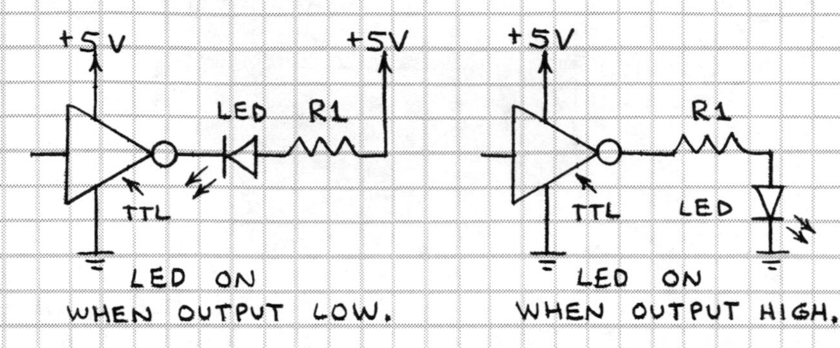


THIS ARRANGEMENT GIVES HIGHER DRIVE CURRENT.

LESS DRIVE CURRENT BUT OK FOR HIGH-BRIGHTNESS LEDS.

R1 CONTROLS DRIVE CURRENT IN BOTH DRIVERS. WHEN $V_{cc}$ = 5 VOLTS AND RED LED IS USED, R1 = 3.3/DESIRED LED CURRENT. EXAMPLE: FOR LED CURRENT OF 10 mA, R = 3.3/.01 = 330Ω.

# PIEZOELECTRIC BUZZER DRIVERS

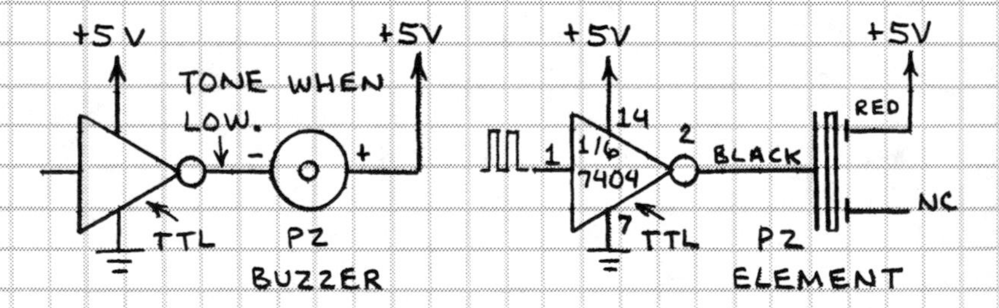


BUZZER DRIVE CURRENT SHOULD NOT EXCEED AVAILABLE OUTPUT CURRENT FROM TTL CHIP.

USE TO CONVERT REPETITIVE INPUT PULSES TO SOUND. ANY TTL INPUT OK.

# TRANSISTOR DRIVERS

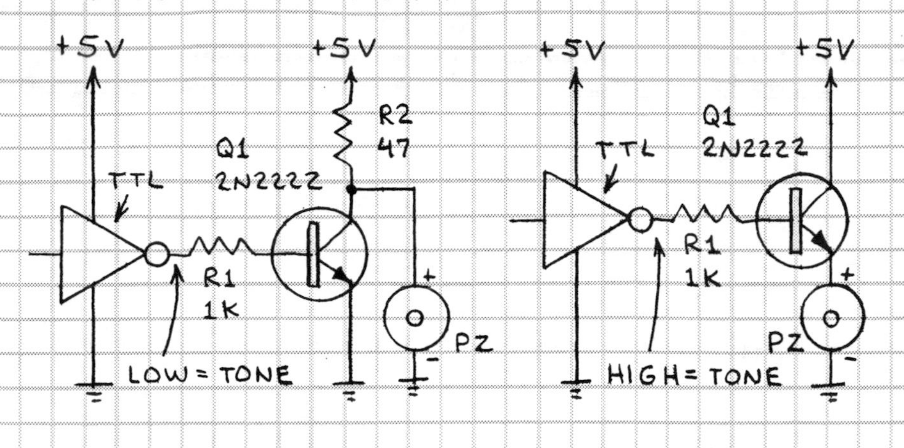


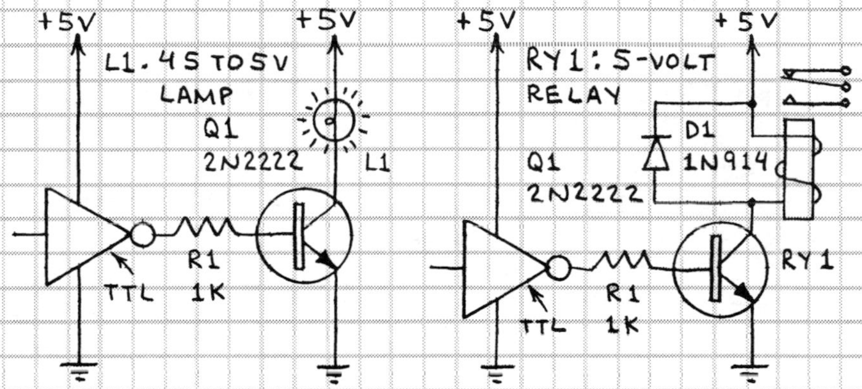


# SCR DRIVERS

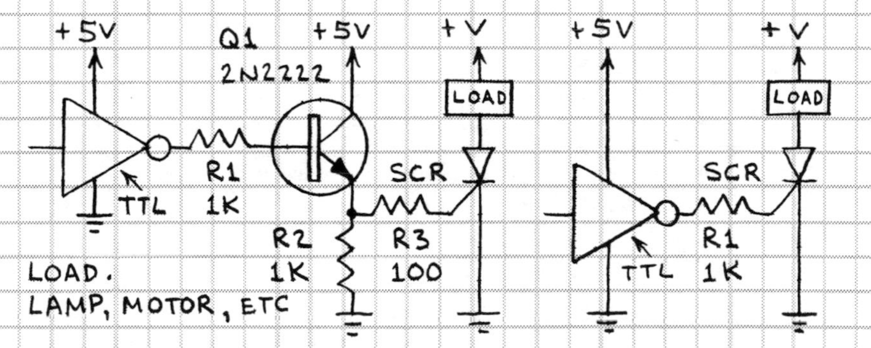


SCR SUPPLY (+V) CAN EXCEED +5V. SCR STAYS ON WHEN TRIGGERED UNLESS FORWARD CURRENT FALLS BELOW SCR HOLDING CURRENT ($I_H$).

# TTL NAND GATE CIRCUITS

USE 7400 OR 7400LS QUAD NAND GATE. PIN NUMBERS ARE GIVEN FOR CONVENIENCE. IF DESIRED, INDIVIDUAL GATES CAN BE REARRANGED.

## CONTROL GATE

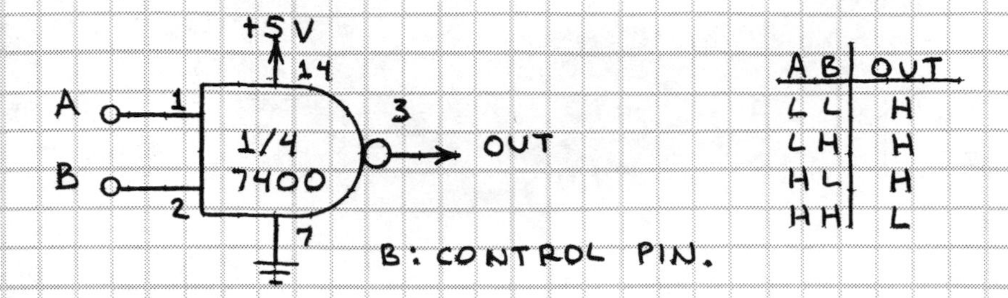


| A B | OUT |
|---|---|
| L L | H |
| L H | H |
| H L | H |
| H H | L |

## AND GATE

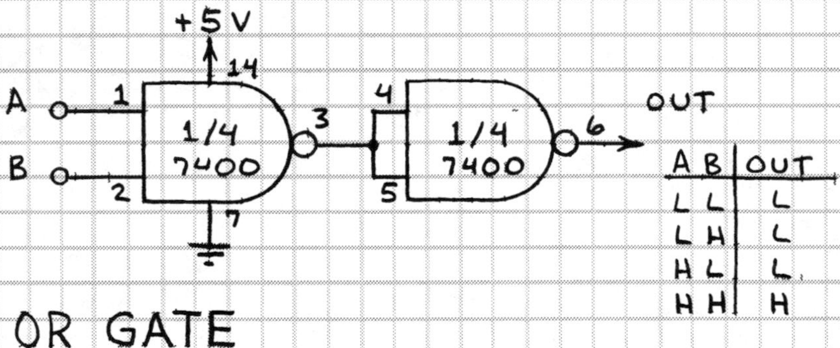


| A B | OUT |
|---|---|
| L L | L |
| L H | L |
| H L | L |
| H H | H |

## OR GATE

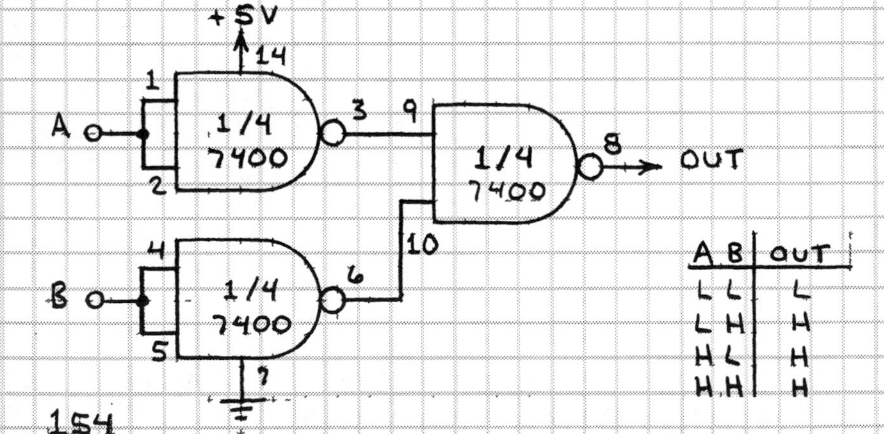


| A B | OUT |
|---|---|
| L L | L |
| L H | H |
| H L | H |
| H H | H |

# INVERTER

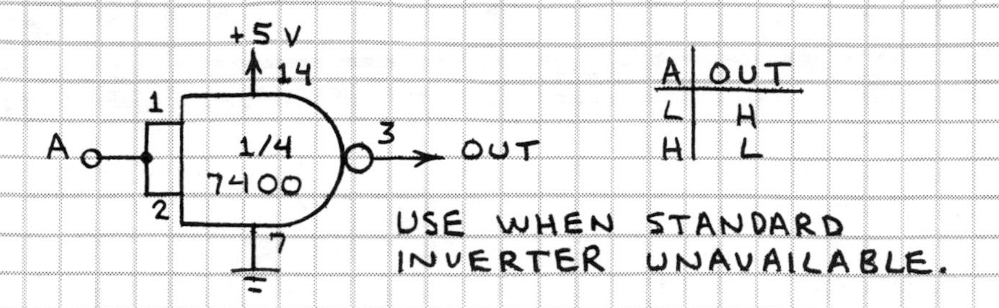


| A | OUT |
|---|---|
| L | H |
| H | L |

# AND-OR GATE

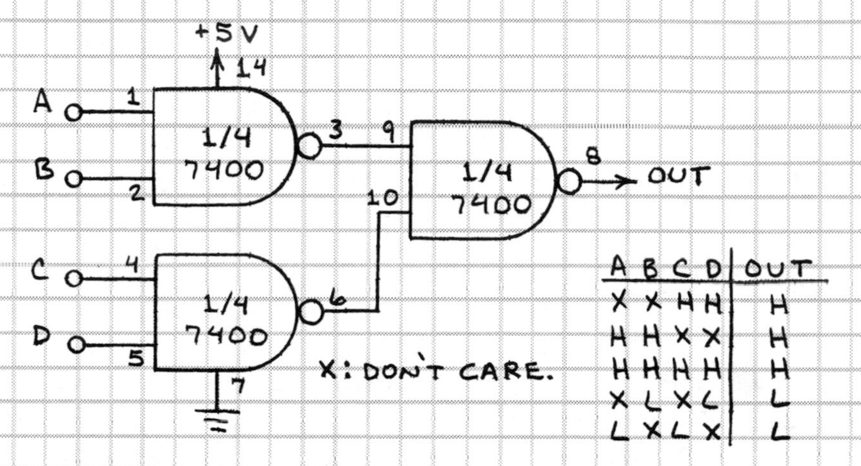


| A | B | C | D | OUT |
|---|---|---|---|---|
| X | X | H | H | H |
| H | H | X | X | H |
| H | H | H | H | H |
| X | L | X | L | L |
| L | X | L | X | L |

# NOR GATE

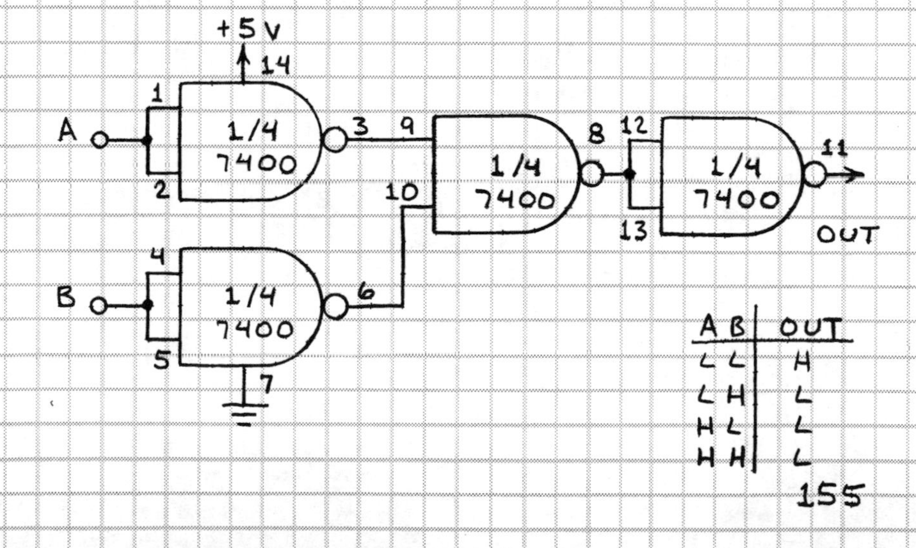


| A | B | OUT |
|---|---|---|
| L | L | H |
| L | H | L |
| H | L | L |
| H | H | L |

# EXCLUSIVE-OR GATE

| A | B | OUT |
|---|---|---|
| L | L | L |
| L | H | H |
| H | L | H |
| H | H | L |

# EXCLUSIVE-NOR GATE

OK TO USE 1/3 7404.

| A | B | OUT |
|---|---|---|
| L | L | H |
| L | H | L |
| H | L | L |
| H | H | H |

# RS LATCH

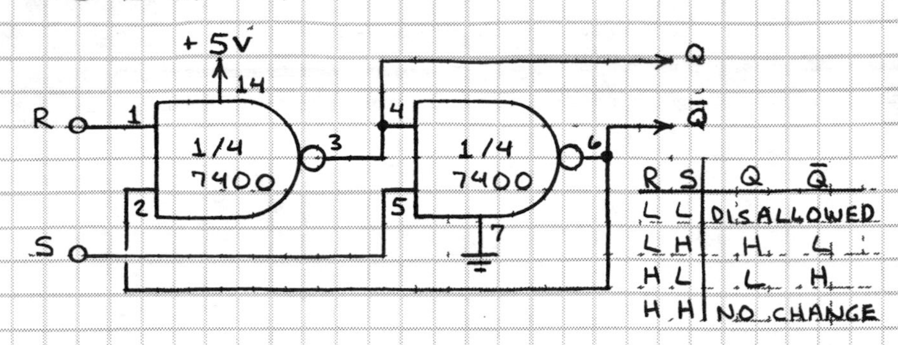


# GATED RS LATCH

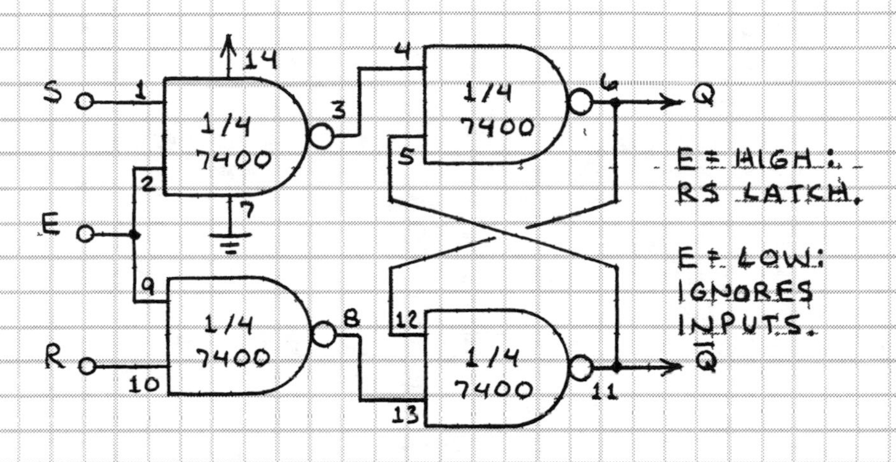


# D FLIP-FLOP

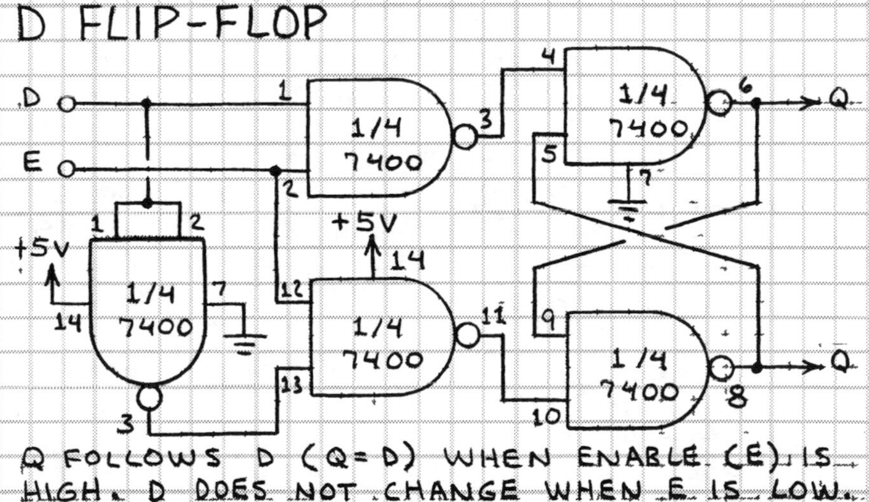


Q FOLLOWS D (Q = D) WHEN ENABLE (E) IS HIGH. D DOES NOT CHANGE WHEN E IS LOW.

# BINARY HALF ADDER

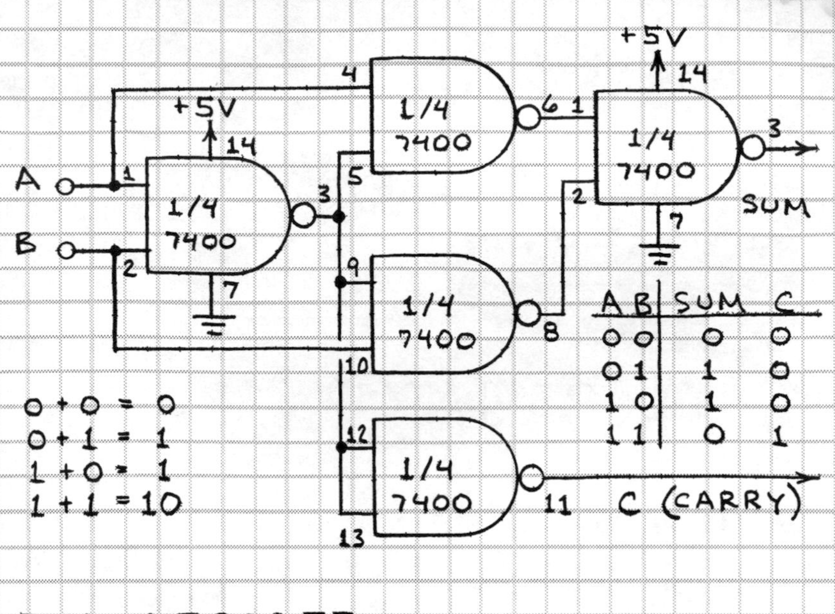


# BCD DECODER

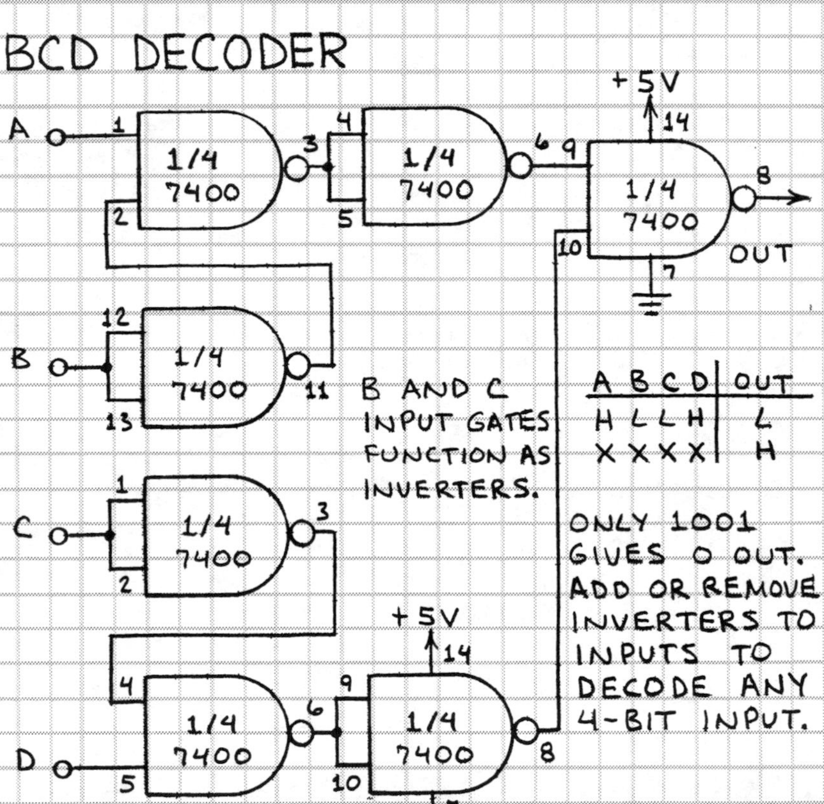

# SWITCH DEBOUNCER

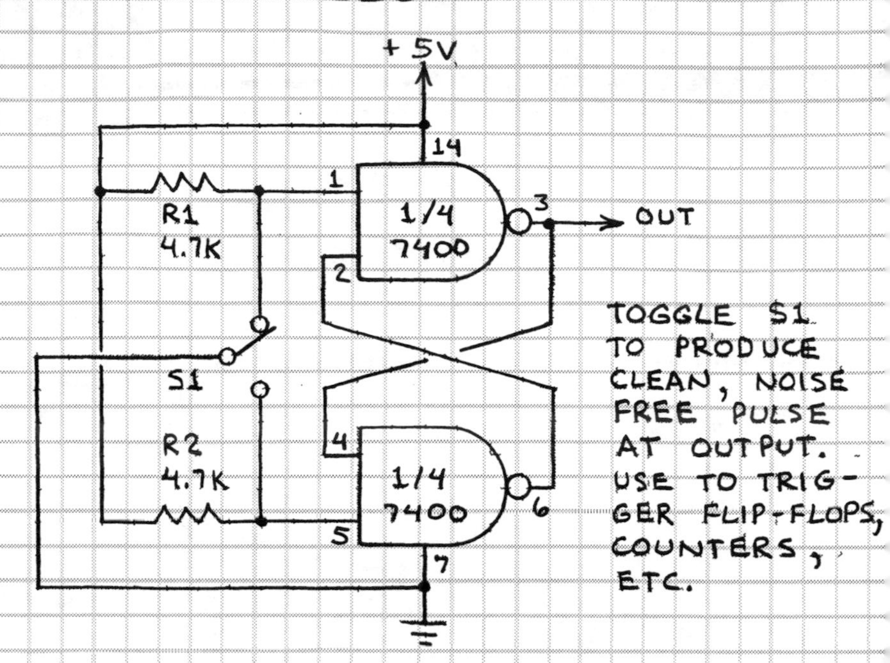


# DUAL LED FLASHER

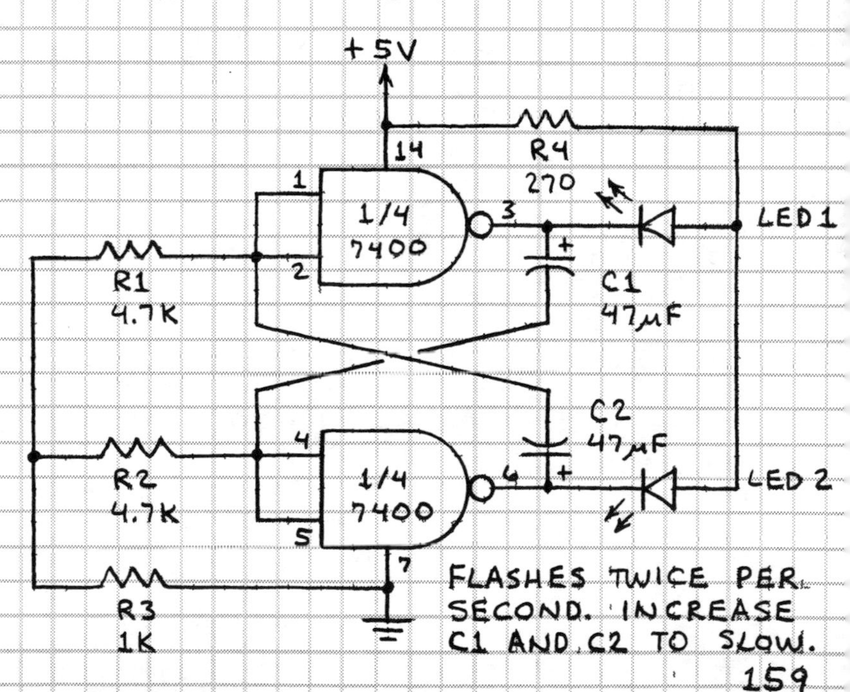

# TTL APPLICATION CIRCUITS

THE CIRCUITS THAT FOLLOW ILLUSTRATE HOW TTL CHIPS CAN BE EASILY INTERCONNECTED TO ACCOMPLISH MANY DIFFERENT APPLICATIONS.

## 1-OF-2 DEMULTIPLEXER

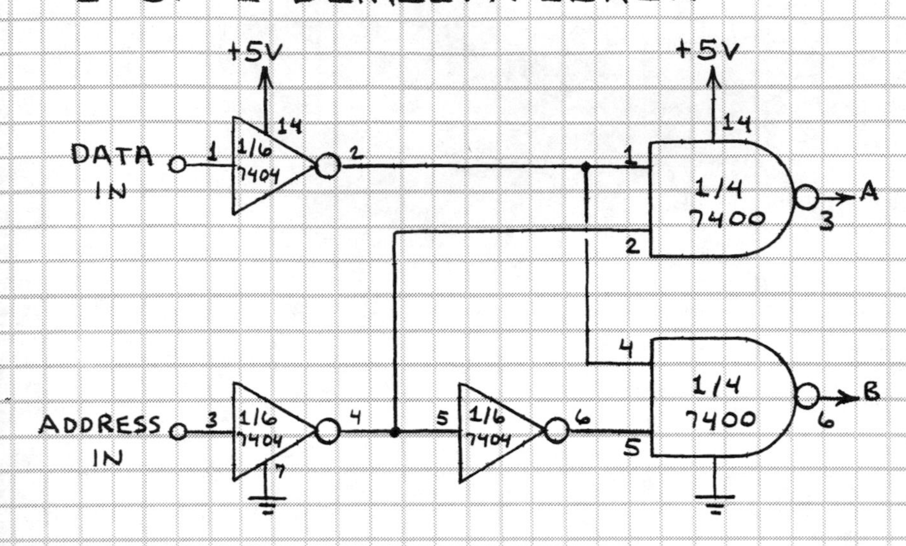


INPUT BIT AT DATA IN IS STEERED TO A OR B OUTPUT BY THE ADDRESS BIT.

| DATA | ADDRESS | A | B |
|---|---|---|---|
| L | L | L | H |
| H | L | H | H |
| L | H | H | L |
| H | H | H | H |

## EXPANDER

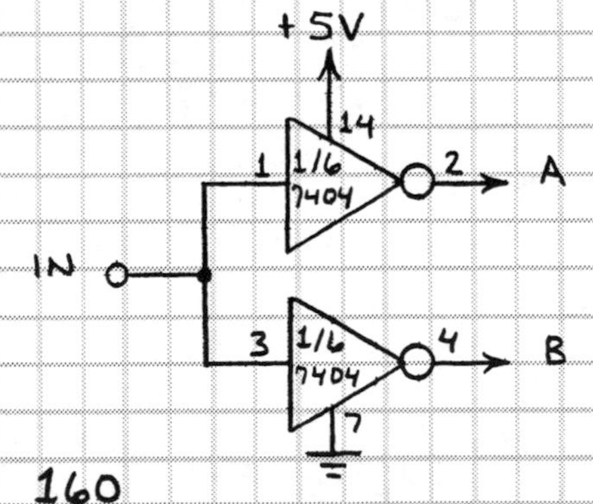


USE TO PROVIDE MULTIPLE OUTPUTS, EACH WITH SAME DRIVE CAPABILITY AS SINGLE OUTPUT. USE FOR LEDS, TRANSISTOR DRIVERS, ETC.

# 2-INPUT DATA SELECTOR

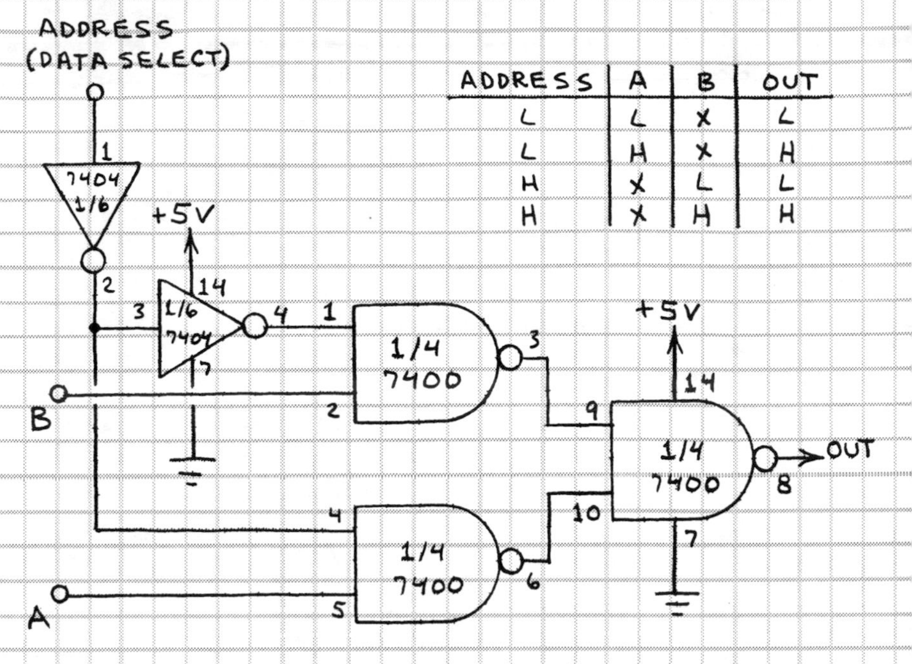


| ADDRESS | A | B | OUT |
|---|---|---|---|
| L | L | X | L |
| L | H | X | H |
| H | X | L | L |
| H | X | H | H |

SELECTED INPUT BIT (A OR B) IS STEERED TO OUTPUT. CIRCUIT CAN BE EXPANDED.

# LOGIC PROBES

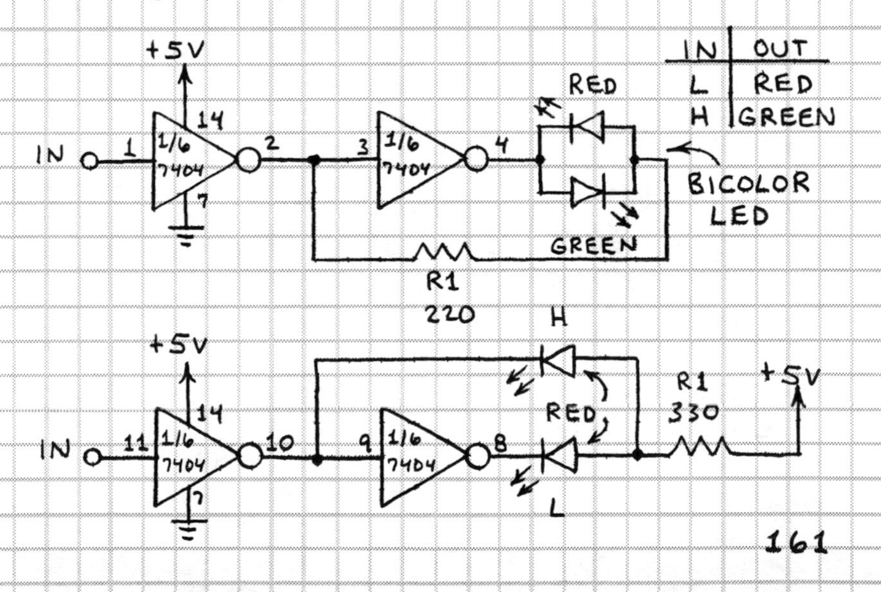


| IN | OUT |
|---|---|
| L | RED |
| H | GREEN |

# UNANIMOUS VOTE DETECTOR

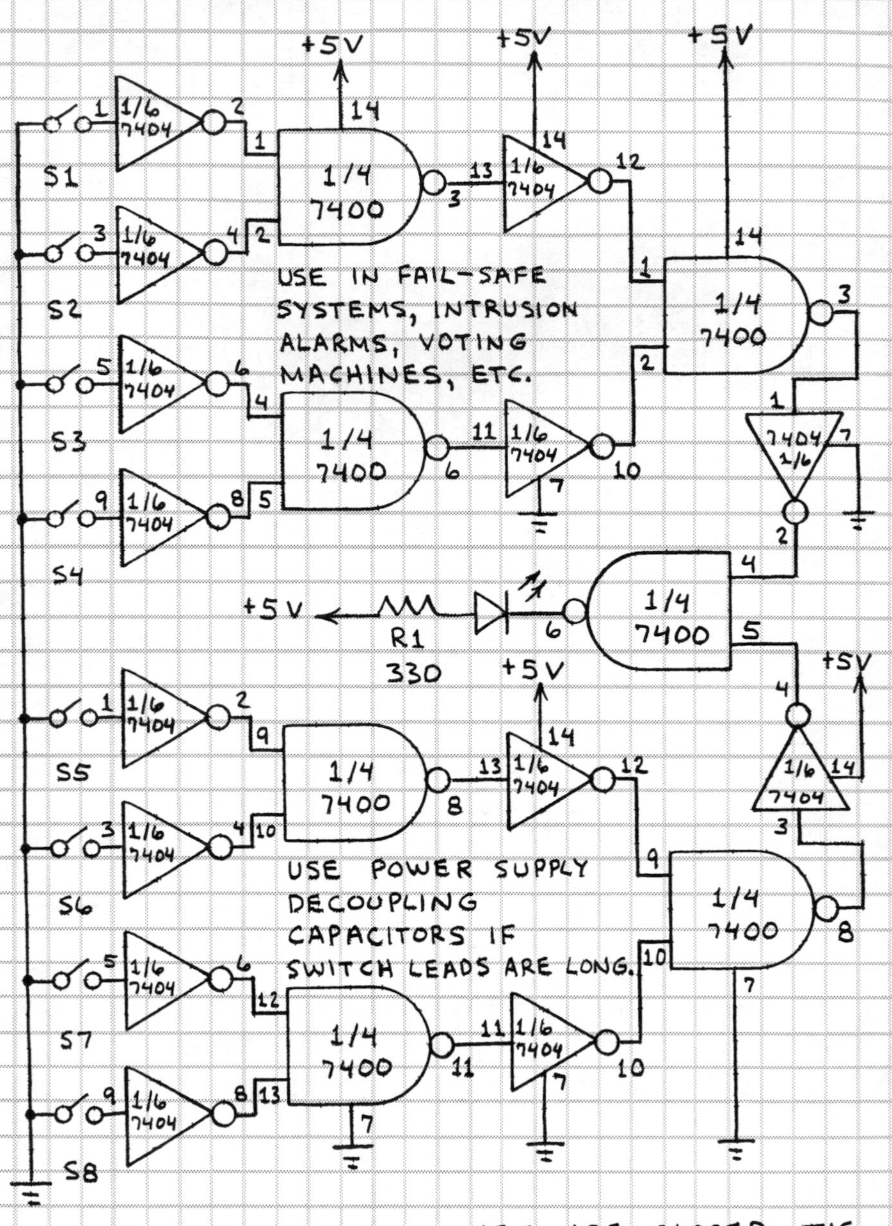


WHEN ALL INPUT SWITCHES ARE CLOSED, THE LED GLOWS. IF OUTPUT IS SENT TO OTHER LOGIC, TIE INPUTS OF 8 7404 INPUT INVERTERS TO +5V THROUGH 4.7K RESISTORS.

# DIVIDE-BY-N COUNTERS

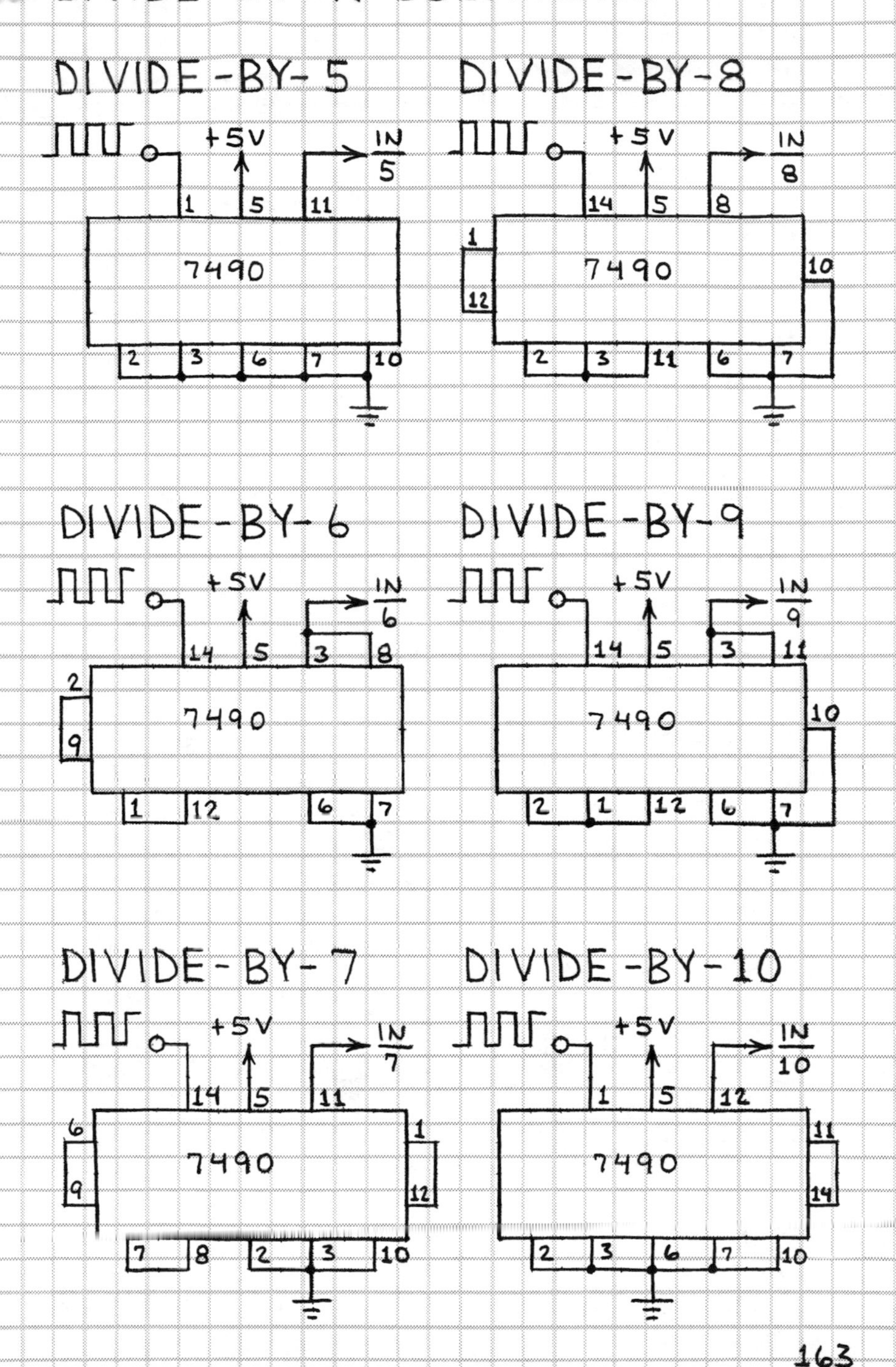

# 2-DIGIT BCD COUNTER

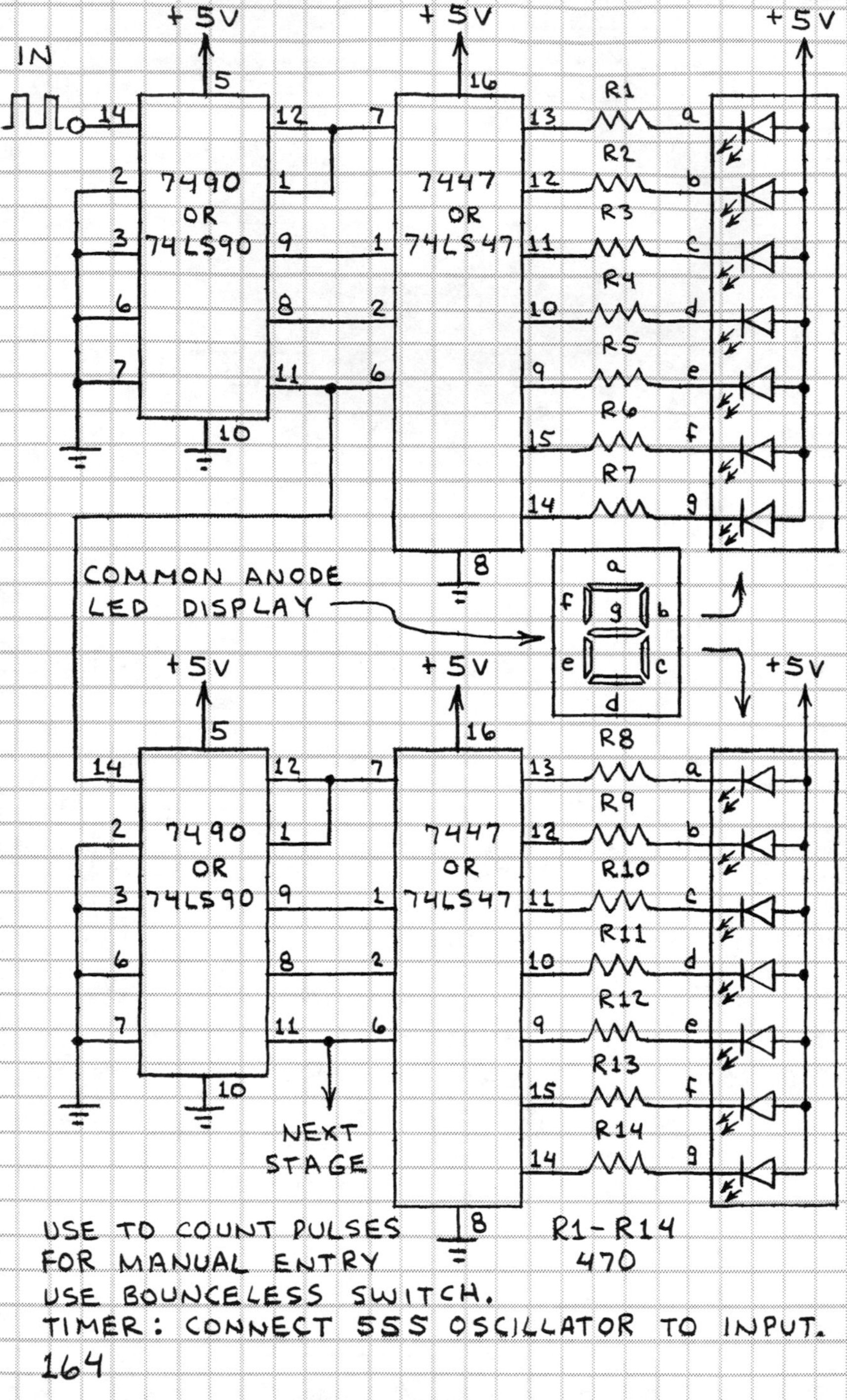

COMMON ANODE LED DISPLAY

R1-R14 470

USE TO COUNT PULSES
FOR MANUAL ENTRY
USE BOUNCELESS SWITCH.
TIMER: CONNECT 555 OSCILLATOR TO INPUT.

# DISPLAY DIMMER / FLASHER

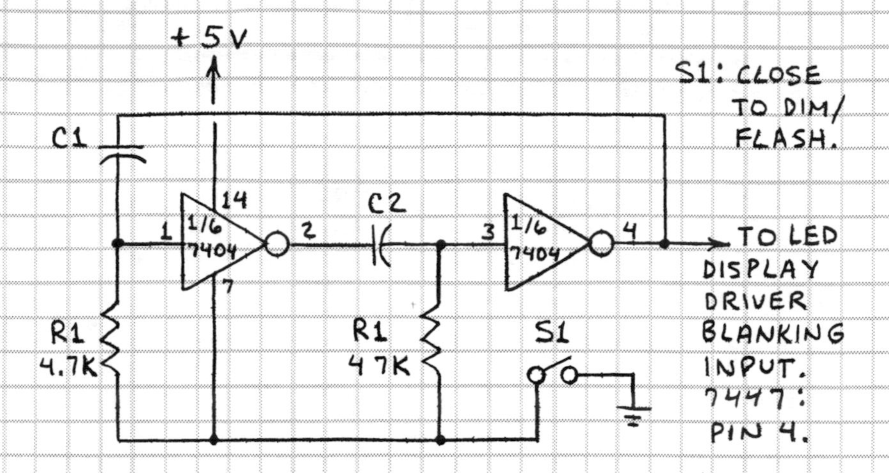


DIMMER : C1, C2 = 0.1 μF
FLASHER: C1, C2 = 47 μF (2 FLASHES PER SECOND)

THIS CIRCUIT WILL CONTROL 7447 DECODERS ON FACING PAGE (CONNECT PIN 4 OF EACH 7447 TO OUTPUT OF DIMMER / FLASHER).

# 0 TO 99 SECOND / MINUTE TIMER

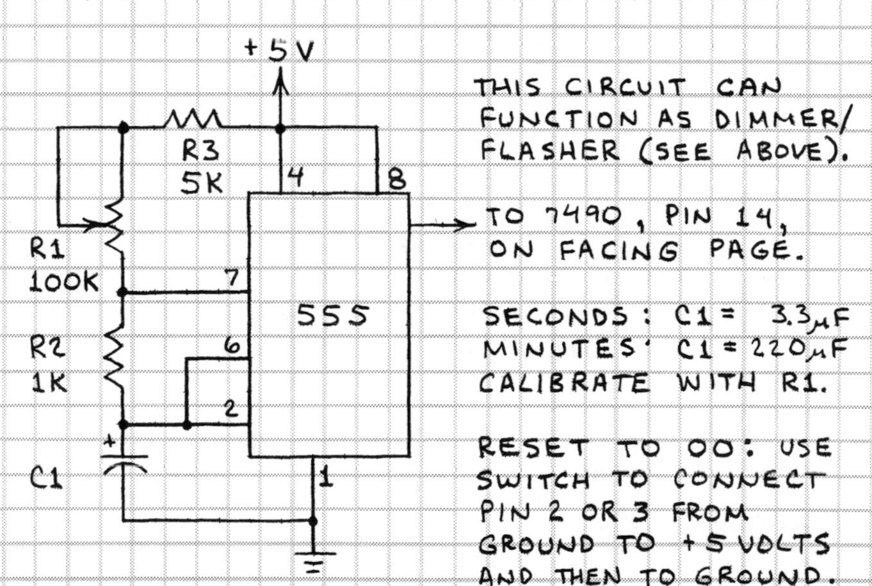


THIS CIRCUIT CAN FUNCTION AS DIMMER/ FLASHER (SEE ABOVE).

SECONDS: C1 = 3.3 μF
MINUTES: C1 = 220 μF
CALIBRATE WITH R1.

RESET TO 00: USE SWITCH TO CONNECT PIN 2 OR 3 FROM GROUND TO +5 VOLTS AND THEN TO GROUND.

# CMOS LOGIC FAMILY

CMOS (COMPLEMENTARY METAL-OXIDE-SILICON) LOGIC CHIPS CAN CONTAIN FAR MORE FUNCTIONS PER CHIP THAN TTL AND TTL/LS LOGIC CHIPS. THOUGH STANDARD CMOS IS NOT AS FAST AS TTL LOGIC, IT CONSUMES CONSIDERABLY LESS POWER. A SINGLE CMOS GATE CONSUMES 0.1 MILLIAMPERE. MOREOVER, CMOS LOGIC CAN BE POWERED BY A WIDE SUPPLY VOLTAGE (3 TO 18 VOLTS). A MAJOR DRAWBACK OF CMOS IS ITS VULNERABILITY TO STATIC ELECTRICITY.

# OPERATING REQUIREMENTS

1 $V_{DD}$ (POSITIVE SUPPLY) MUST <u>NOT</u> EXCEED 15 VOLTS (STANDARD CMOS) OR 18 VOLTS (B SERIES).

2. INPUT SIGNAL MUST <u>NEVER</u> EXCEED $V_{DD}$ NOR FALL BELOW GROUND.

3. UNUSED INPUTS <u>WILL</u> PICK UP STRAY SIGNALS AND CAUSE ERRATIC OPERATION AND EXCESSIVE POWER CONSUMPTION. <u>ALL</u> UNUSED INPUTS <u>MUST</u> BE CONNECTED TO $V_{DD}$ OR GROUND.

4. IF POSSIBLE, AVOID INPUT SIGNALS THAT CHANGE STATES SLOWLY SINCE THEY INCREASE POWER CONSUMPTION. RISE AND FALL TIMES FASTER THAN 15 MICROSECONDS ARE BEST.

5. THE FREQUENCY OF THE INPUT SIGNAL MUST NOT EXCEED THE MAXIMUM OPERATING FREQUENCY OF A CMOS CHIP. A STANDARD CMOS CHIP HAS A TYPICAL MAXIMUM RESPONSE OF 1 MHz WHEN $V_{DD}$ = 5 VOLTS AND 5 MHz WHEN $V_{DD}$ = 15 VOLTS.

6. <u>NEVER</u> CONNECT AN INPUT SIGNAL TO A CMOS CHIP WHEN THE POWER IS OFF. <u>NEVER</u> REMOVE POWER TO A CMOS CHIP WHEN AN INPUT SIGNAL IS PRESENT.

# HANDLING PRECAUTIONS

1. AVOID TOUCHING THE PINS OF CMOS CHIPS.

2. NEVER STORE CMOS CHIPS IN NON-CONDUCTIVE PLASTIC TRAYS, BAGS, FOAM, OR "SNOW."

3. PLACE CMOS CHIPS PINS DOWN ON AN ALUMINUM FOIL SHEET OR TRAY WHEN THEY ARE NOT IN A CIRCUIT OR STORED IN CONDUCTIVE FOAM.

4. NEVER INSTALL A CMOS CHIP IN A CIRCUIT WHEN POWER IS APPLIED. NEVER REMOVE A CMOS CHIP FROM A CIRCUIT WHEN POWER IS APPLIED.

5. USE A BATTERY-POWERED IRON TO MAKE SOLDER CONNECTIONS TO A CMOS CHIP. AN AC POWERED IRON MAY BE USED IF THE TIP DOES NOT CARRY STRAY VOLTAGE.

# POWER SUPPLIES

MOST CMOS CIRCUITS CAN BE POWERED BY BATTERIES. GENERALLY, OUTPUT DEVICES LIKE LEDS, LAMPS, RELAYS, ETC. CONSUME MUCH MORE POWER THAN THE CMOS CHIPS THAT DRIVE THEM.

# BATTERY POWER SUPPLIES

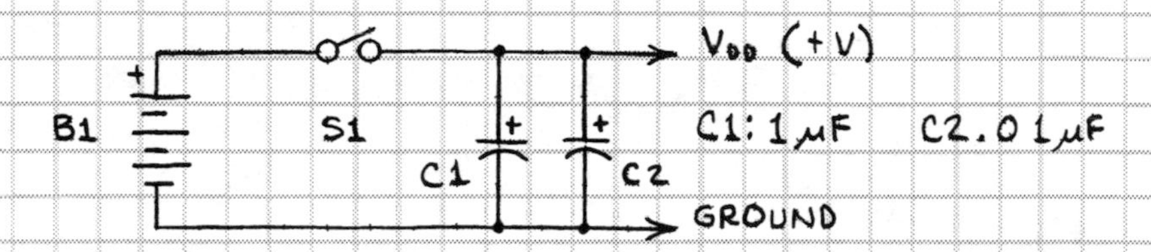


B1 IS 3 TO 15 VOLT BATTERY. C1 AND C2 ARE OPTIONAL. USE WHEN LEADS TO B1 ARE LONG. OK TO USE 7805, 7812, OR 7815 REGULATOR CHIP IN BATTERY SUPPLY ON P 11.

# CMOS INPUT INTERFACING

NON-CMOS CHIPS AND COMPONENTS CAN SUPPLY INPUT SIGNALS TO CMOS CHIPS IF THE OPERATING REQUIREMENTS ON PAGE 28 ARE OBSERVED. THE FINAL INVERTER IN EACH CIRCUIT BELOW REPRESENTS A CMOS INPUT.

# CLOCK PULSE GENERATORS

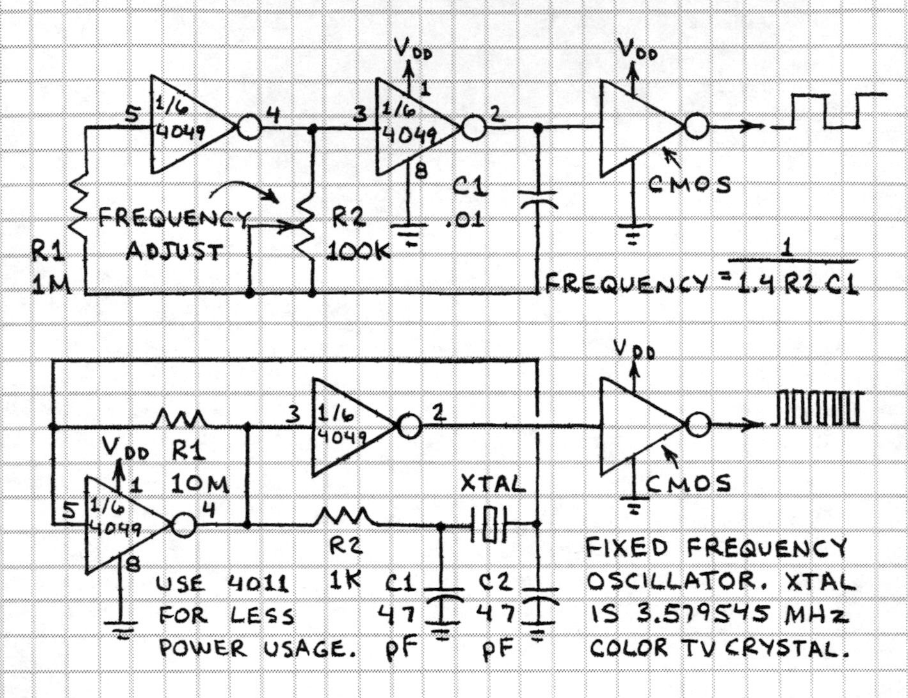


# BOUNCELESS SWITCH

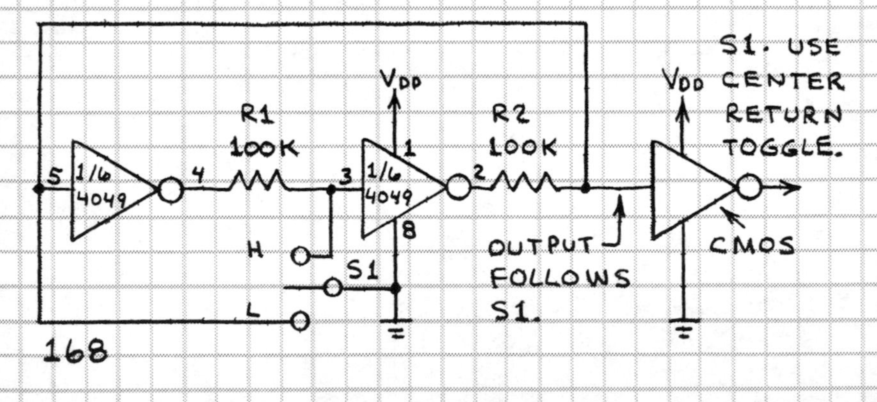

# PHOTOCELL TO CMOS

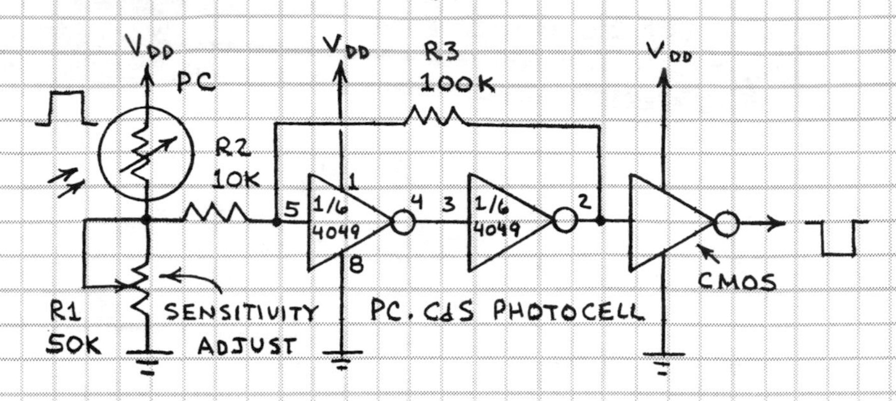


# PHOTOTRANSISTOR TO CMOS

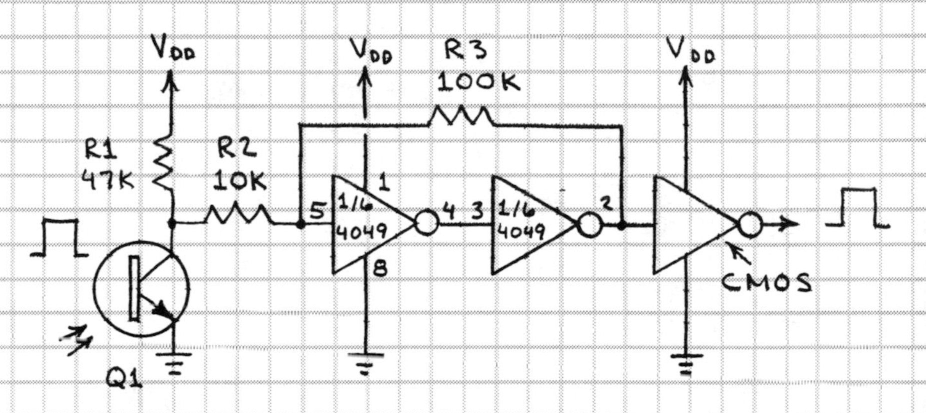


# COMPARATOR / OP-AMP TO CMOS

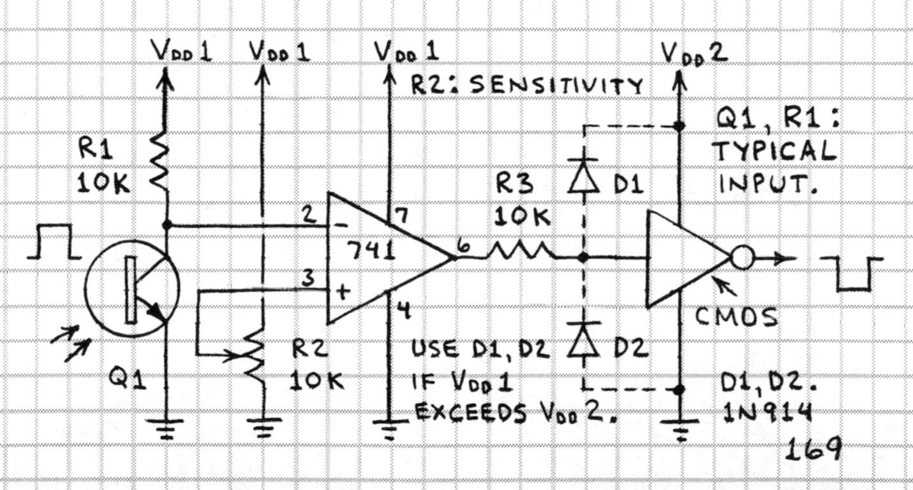

# CMOS OUTPUT INTERFACING

THOUGH CMOS CHIPS HAVE LIMITED OUTPUT CURRENT, MANY OUTPUT DEVICES CAN BE DRIVEN WITH THE HELP OF EXTERNAL COMPONENTS.

# INCREASED OUTPUT

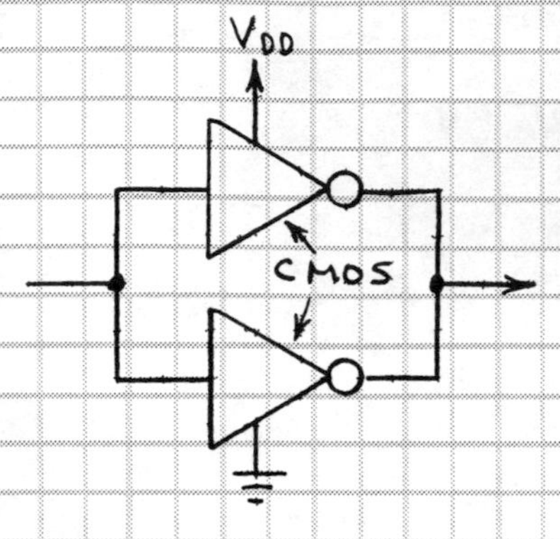


CONNECT TWO OR MORE GATES IN PARALLEL TO INCREASE OUTPUT CURRENT. TWO GATES SHOWN HAVE ABOUT DOUBLE THE OUTPUT AS A SINGLE GATE. THE 4049 AND 4050 HEX INVERTER AND BUFFER GIVE HIGH OUTPUT.

# LED DRIVERS

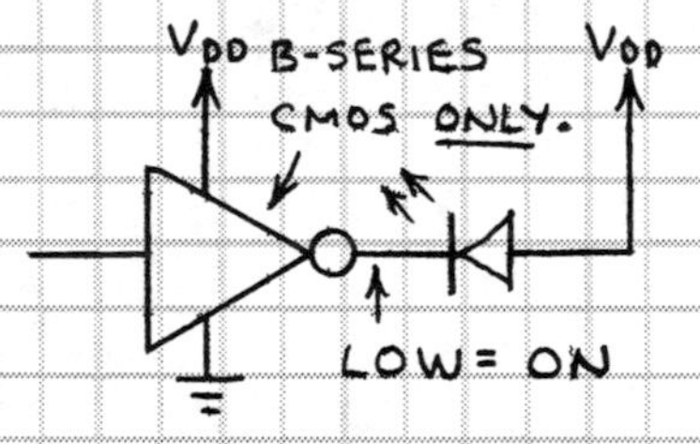


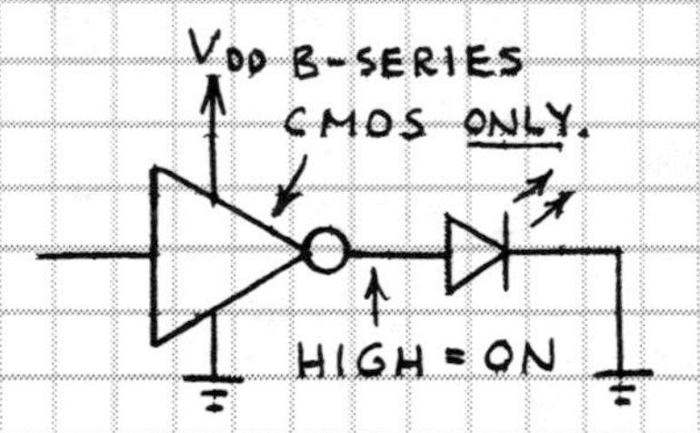


LED WITHOUT RESISTOR FOR $V_{DD} \leq 4.5$ VOLTS ONLY.

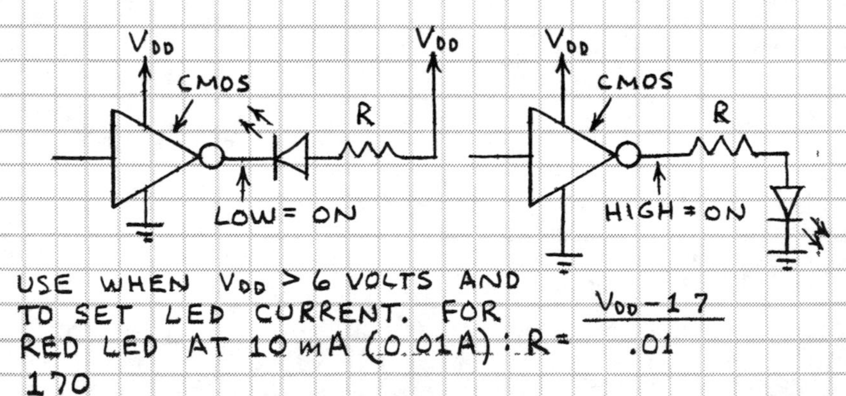


USE WHEN $V_{DD} > 6$ VOLTS AND TO SET LED CURRENT. FOR RED LED AT 10 mA (0.01A): $R = \dfrac{V_{DD} - 1.7}{.01}$

# TRANSISTOR DRIVERS

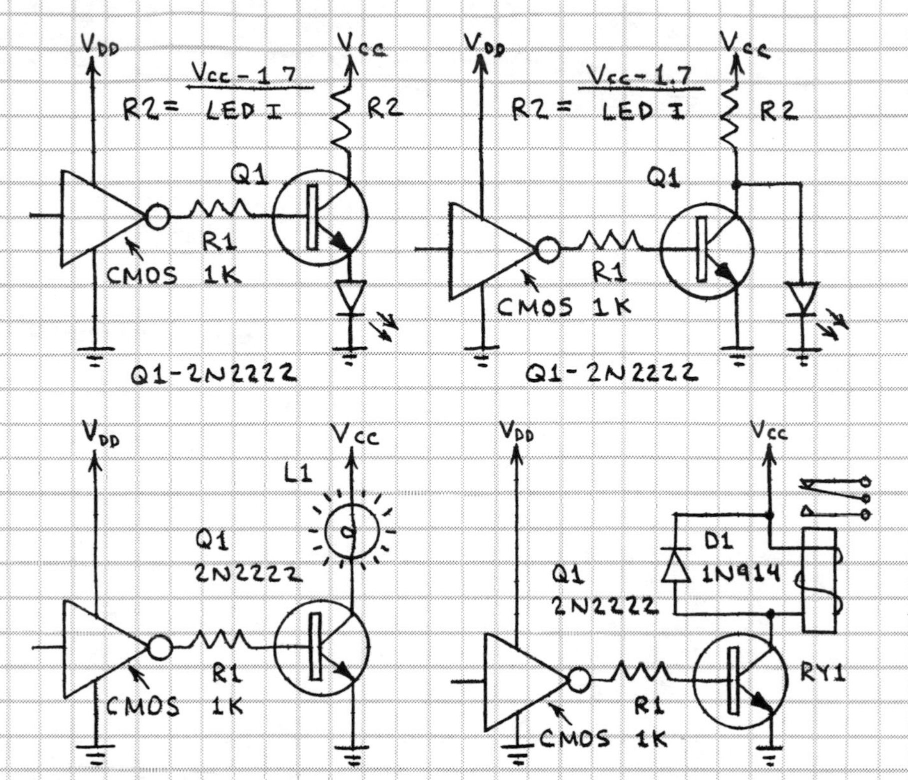


$V_{CC}$ CAN BE $>$ OR $<$ $V_{DD}$. SELECT L1 AND RY ACCORDING TO $V_{CC}$.

# SCR DRIVERS

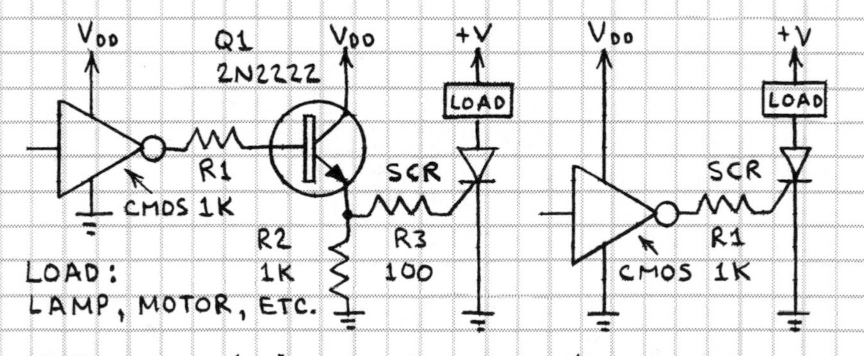


SCR SUPPLY (+V) CAN BE $>$ OR $<$ $V_{DD}$. THESE CIRCUITS IDENTICAL TO TTL VERSIONS ON P. 15.

# CMOS NAND GATE CIRCUITS

USE 4011 QUAD NAND GATE. OK TO REARRANGE GATES. ALL UNUSED INPUTS MUST GO TO $V_{DD}$ OR GROUND. $V_{DD}$ = +3 TO +15 VOLTS. FOLLOW CMOS HANDLING PRECAUTIONS.

## CONTROL GATE

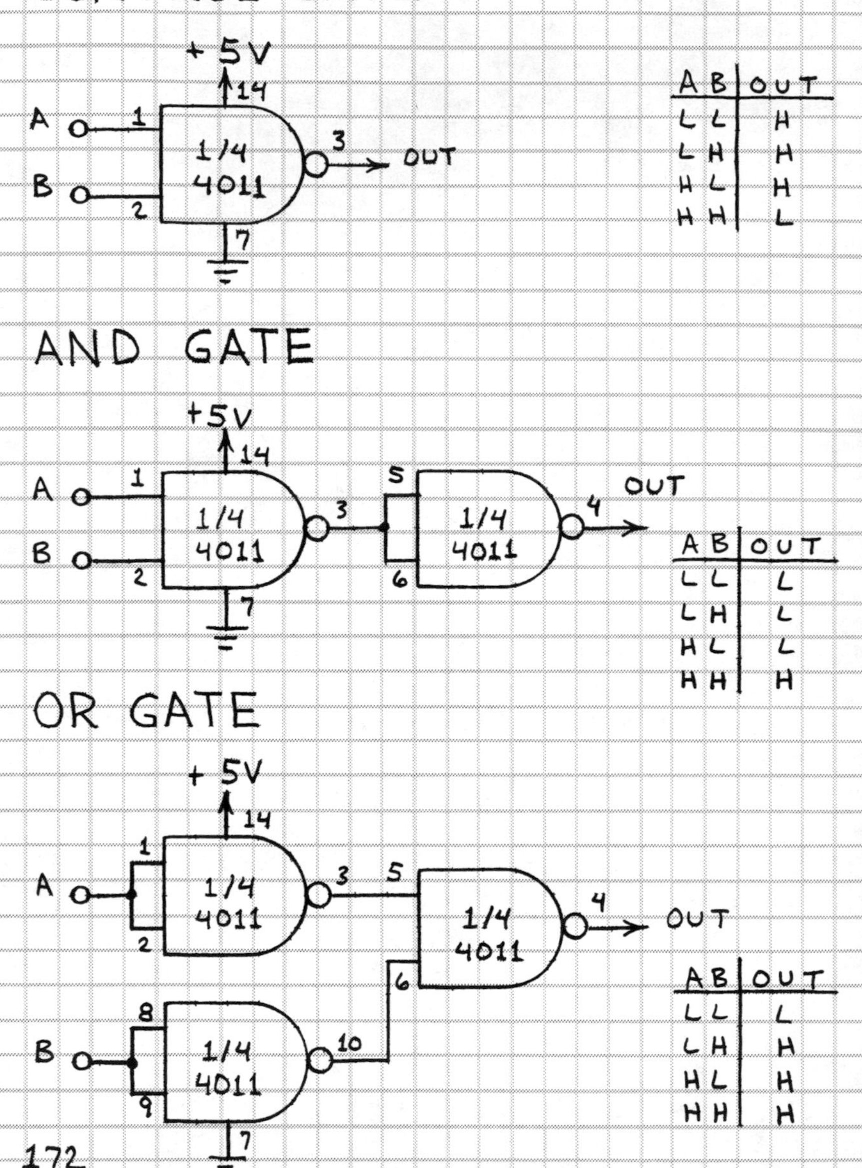


Control gate:

| A | B | OUT |
|---|---|---|
| L | L | H |
| L | H | H |
| H | L | H |
| H | H | L |

AND gate:

| A | B | OUT |
|---|---|---|
| L | L | L |
| L | H | L |
| H | L | L |
| H | H | H |

OR gate:

| A | B | OUT |
|---|---|---|
| L | L | L |
| L | H | H |
| H | L | H |
| H | H | H |

# 4-INPUT NAND GATE

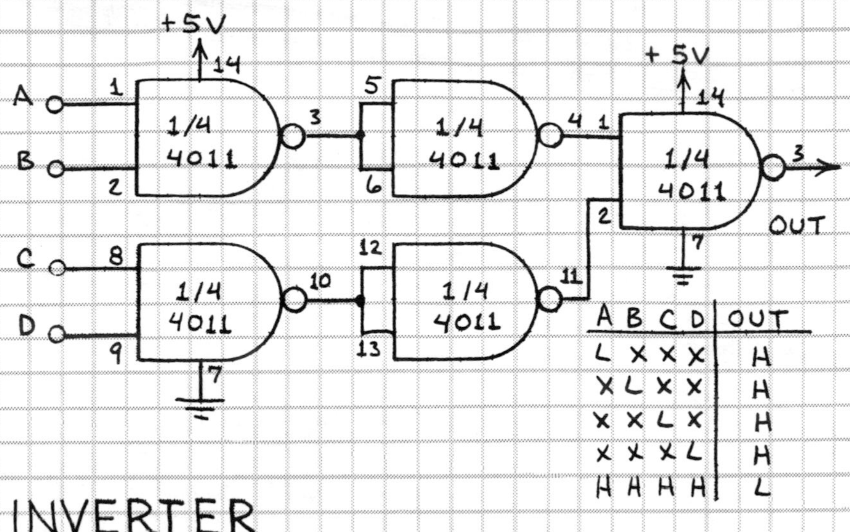


| A | B | C | D | OUT |
|---|---|---|---|---|
| L | X | X | X | H |
| X | L | X | X | H |
| X | X | L | X | H |
| X | X | X | L | H |
| H | H | H | H | L |

# INVERTER

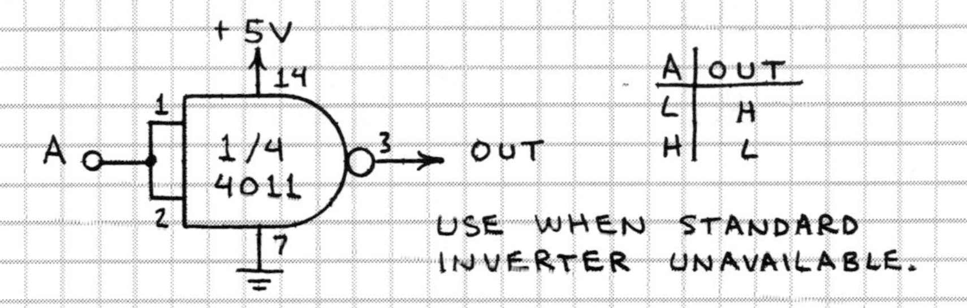


| A | OUT |
|---|---|
| L | H |
| H | L |

USE WHEN STANDARD INVERTER UNAVAILABLE.

# NOR GATE

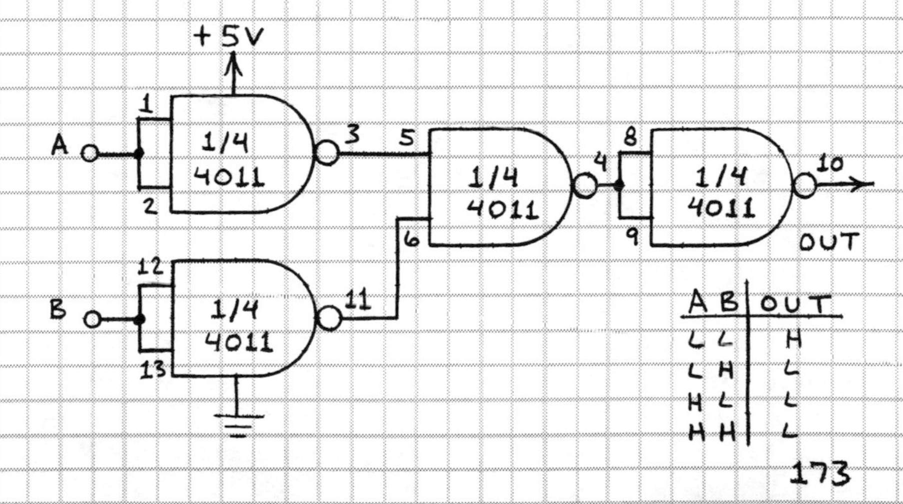


| A | B | OUT |
|---|---|---|
| L | L | H |
| L | H | L |
| H | L | L |
| H | H | L |

# SWITCH DEBOUNCER

TOGGLE S1 TO PRODUCE CLEAN, NOISE FREE OUTPUT PULSE.

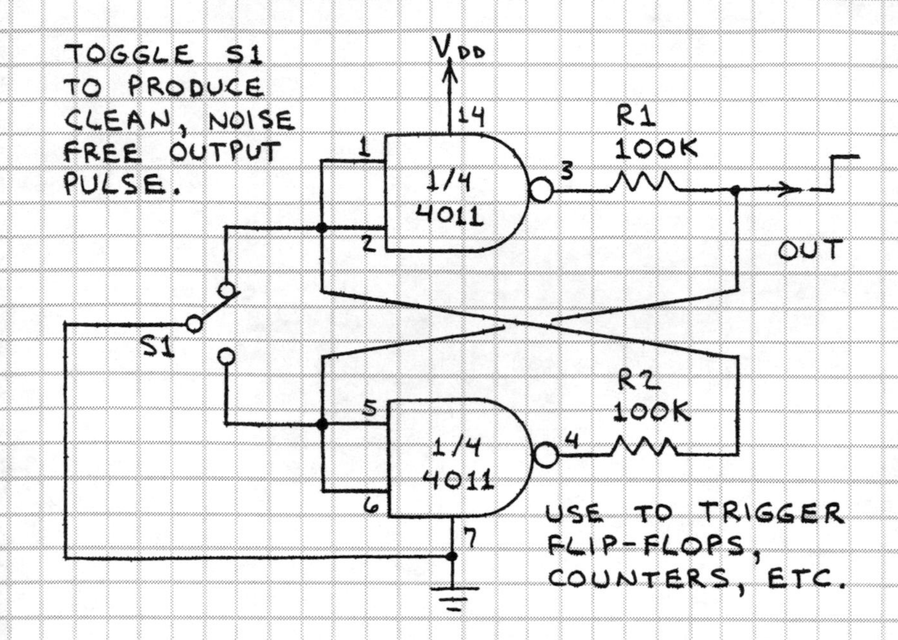


USE TO TRIGGER FLIP-FLOPS, COUNTERS, ETC.

# ONE-SHOT TOUCH SWITCH

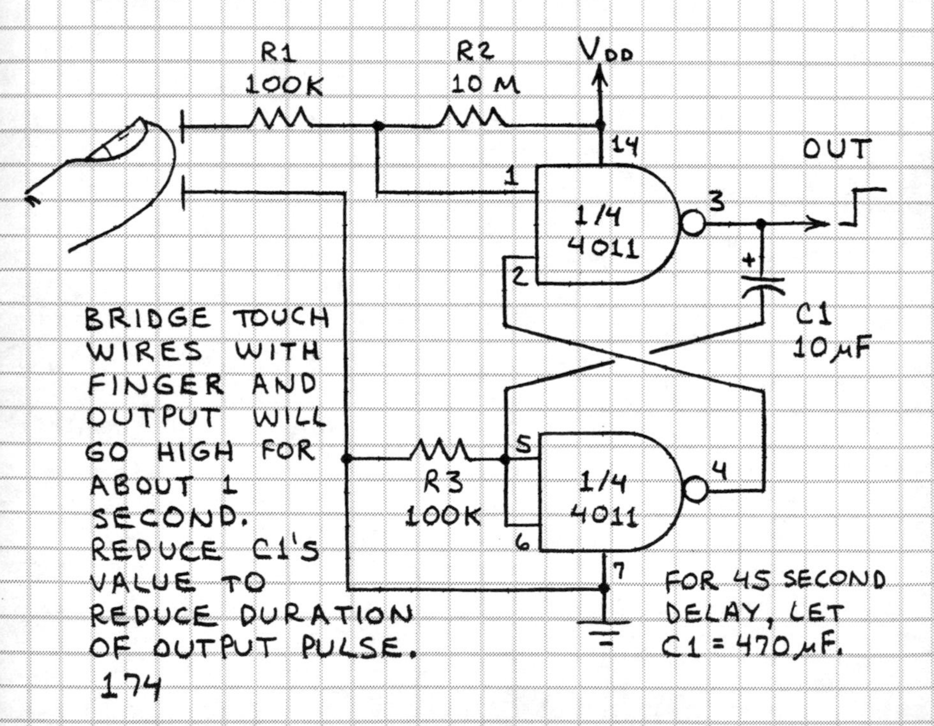


BRIDGE TOUCH WIRES WITH FINGER AND OUTPUT WILL GO HIGH FOR ABOUT 1 SECOND. REDUCE C1'S VALUE TO REDUCE DURATION OF OUTPUT PULSE.

FOR 45 SECOND DELAY, LET C1 = 470μF.

# STANDARD TOUCH SWITCH

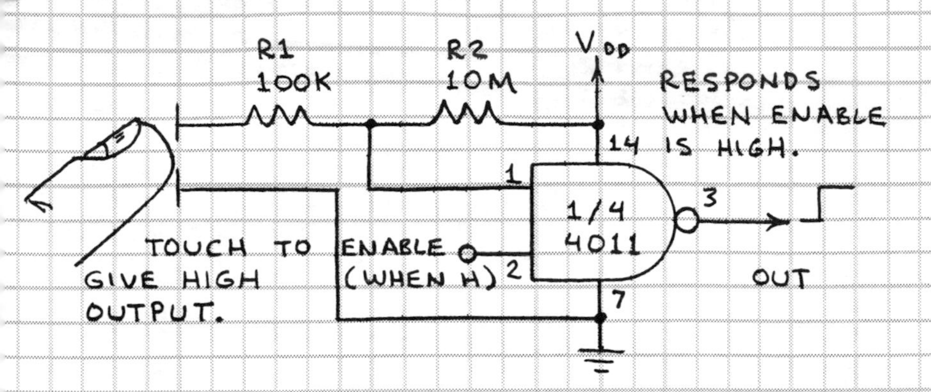


# X-10 LINEAR AMPLIFIER

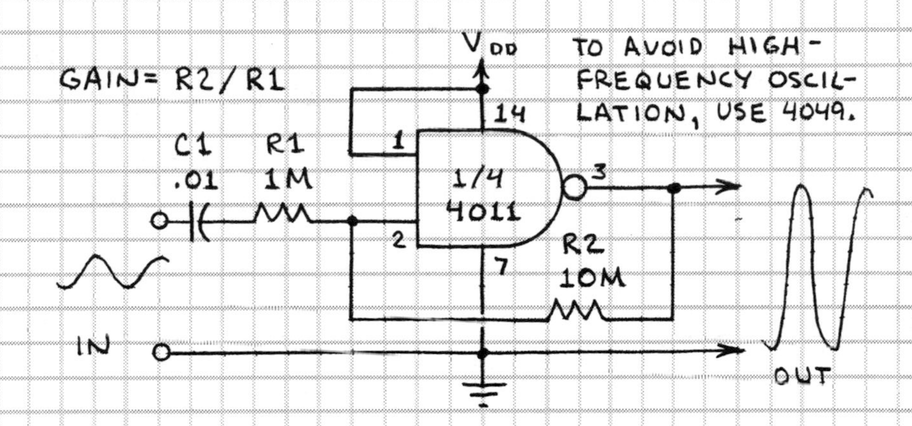


# LAMP FLASHER

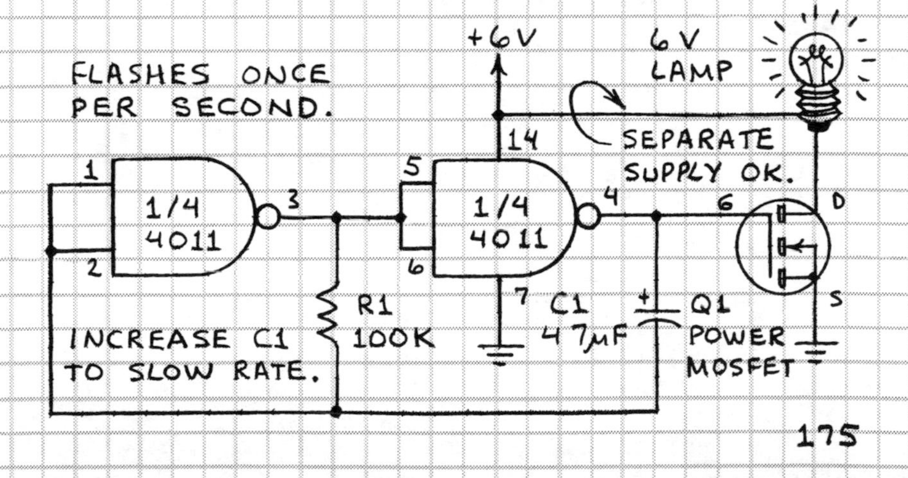

# SIMPLE OSCILLATOR

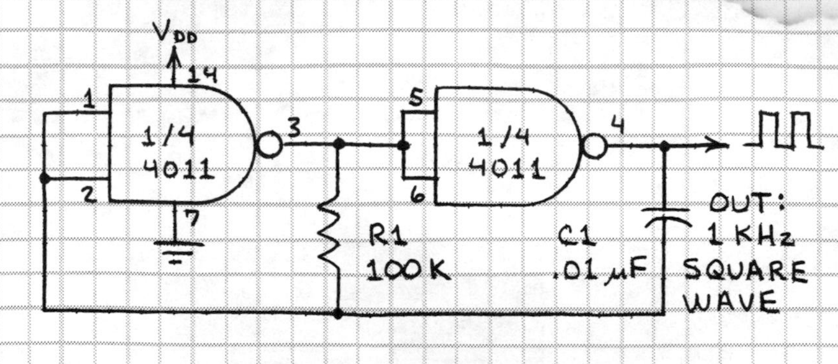


# GATED OSCILLATOR

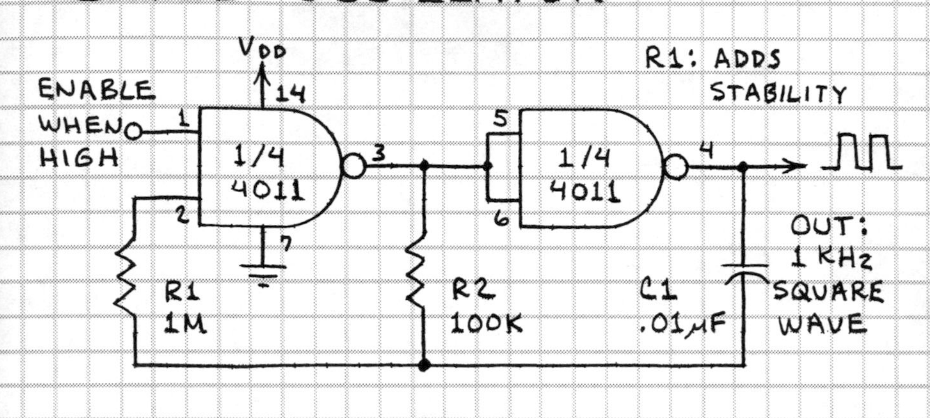


# GATED LED FLASHER

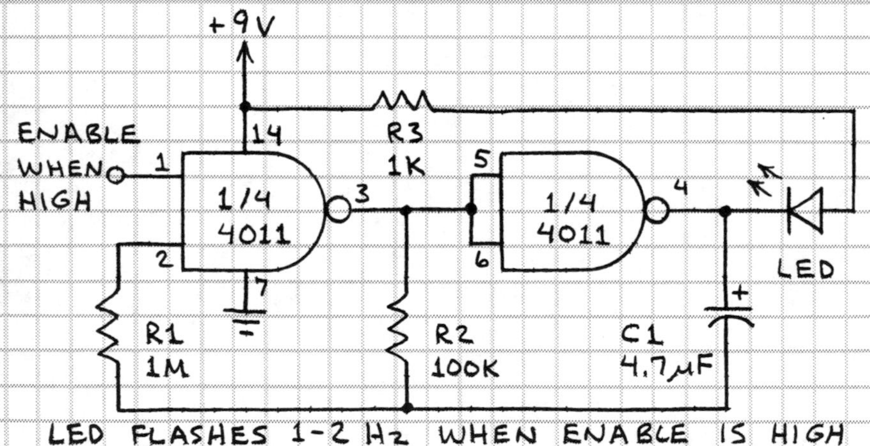


LED FLASHES 1-2 Hz WHEN ENABLE IS HIGH AND GLOWS CONTINUALLY WHEN ENABLE IS LOW.

# GATED TONE GENERATOR

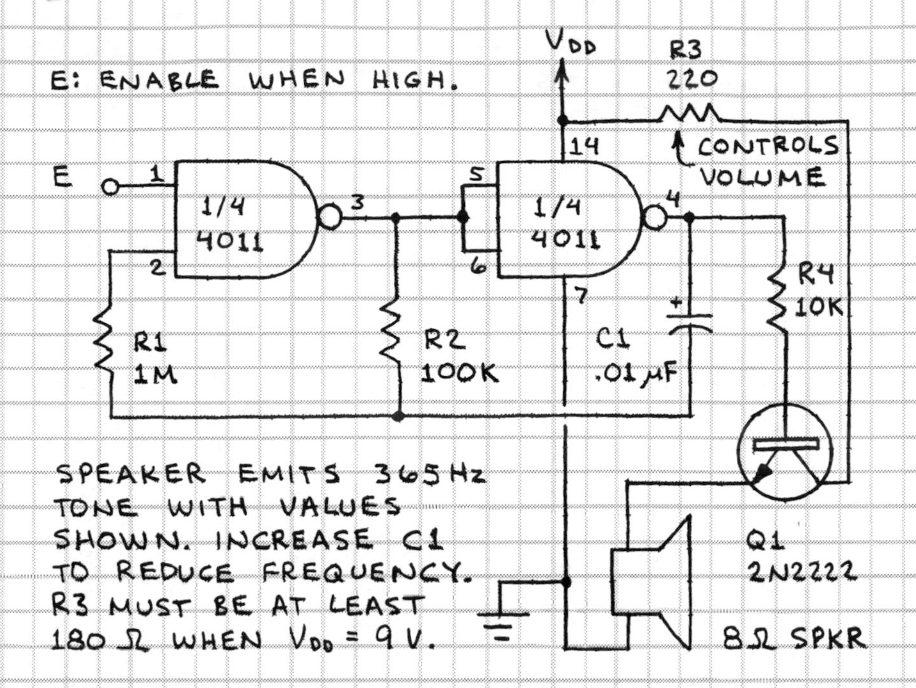


# DUAL LED FLASHER

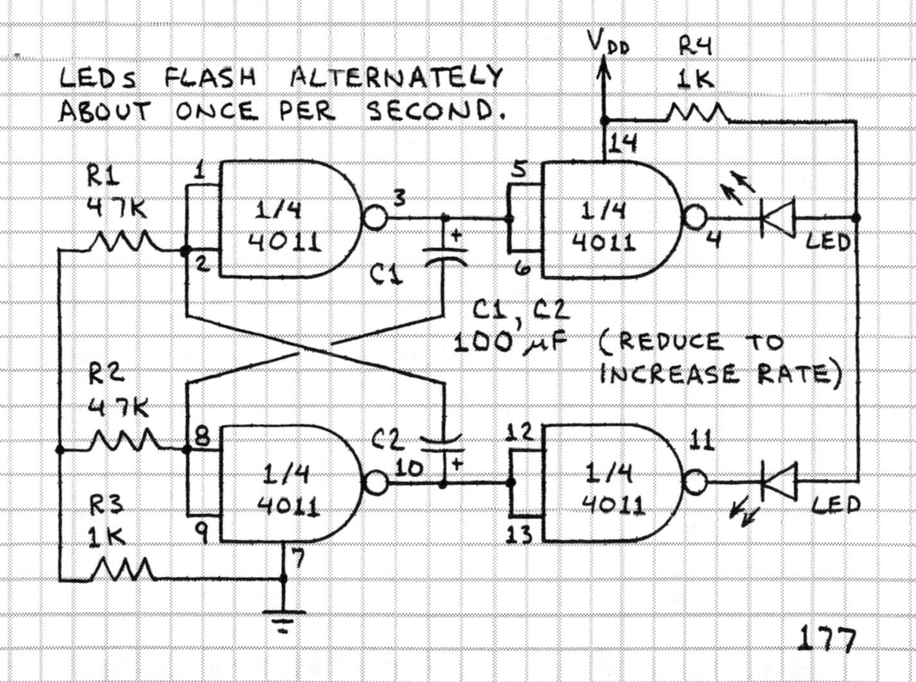

# CMOS APPLICATION CIRCUITS

THE FOLLOWING CIRCUITS ILLUSTRATE THE VERSATILITY OF CMOS LOGIC CHIPS. <u>ALL</u> UNUSED INPUT PINS <u>MUST</u> GO TO $V_{DD}$ OR GROUND.

## RS LATCH

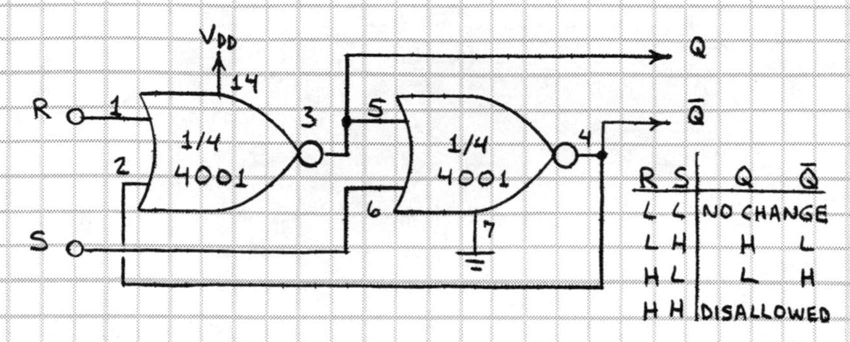


| R | S | Q | Q̄ |
|---|---|---|---|
| L | L | NO CHANGE | |
| L | H | H | L |
| H | L | L | H |
| H | H | DISALLOWED | |

## PHASE-SHIFT OSCILLATOR

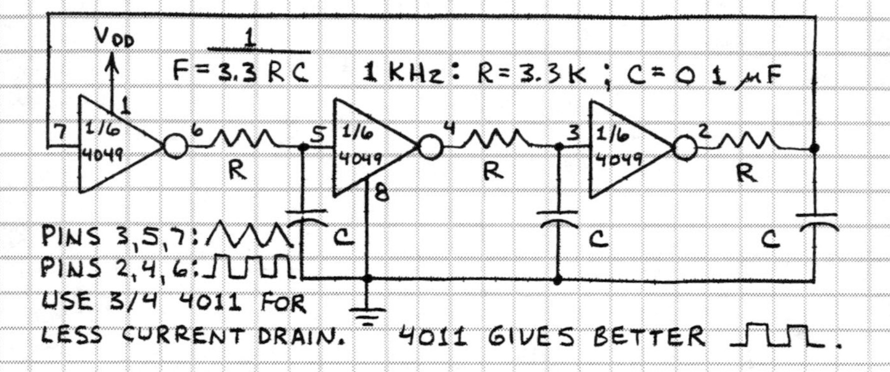


## LOGIC PROBE

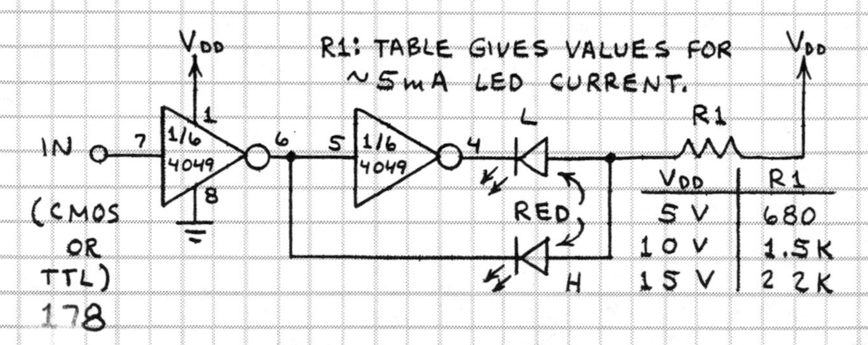


| VDD | R1 |
|---|---|
| 5 V | 680 |
| 10 V | 1.5K |
| 15 V | 2 2K |

# 4-BIT DATA BUS CONTROL

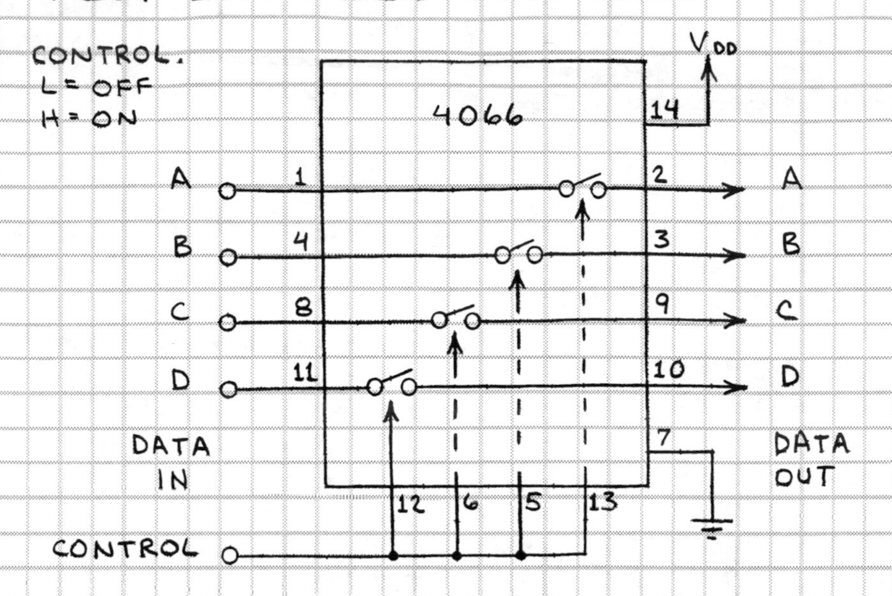


# 1-OF-4 DATA SELECTOR

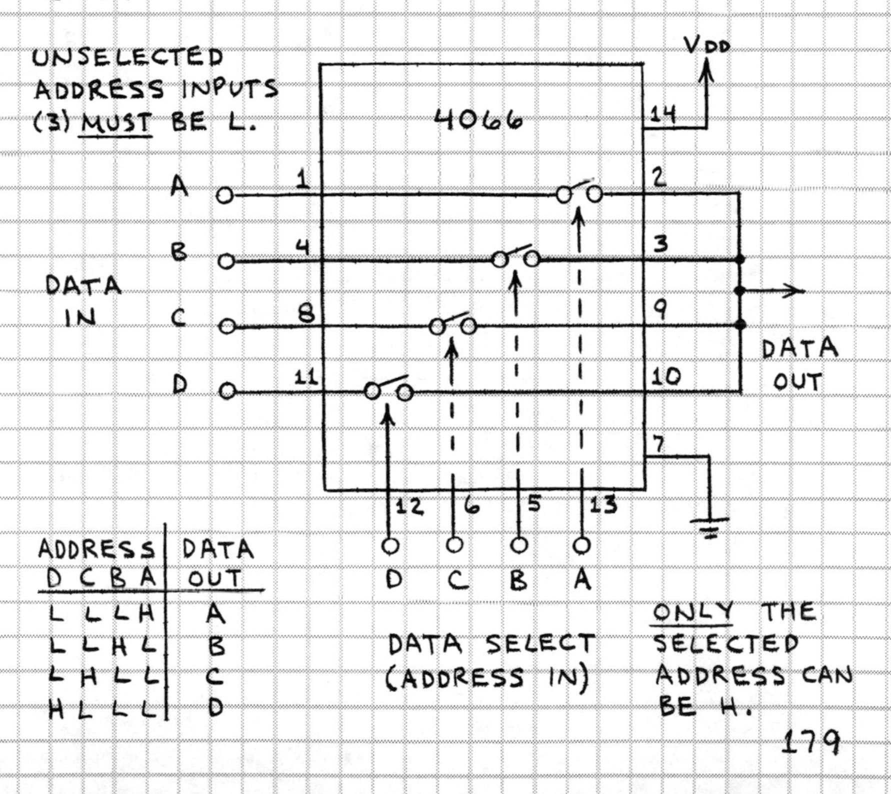


| ADDRESS D C B A | DATA OUT |
|---|---|
| L L L H | A |
| L L H L | B |
| L H L L | C |
| H L L L | D |

# 1-OF-4 SEQUENCER

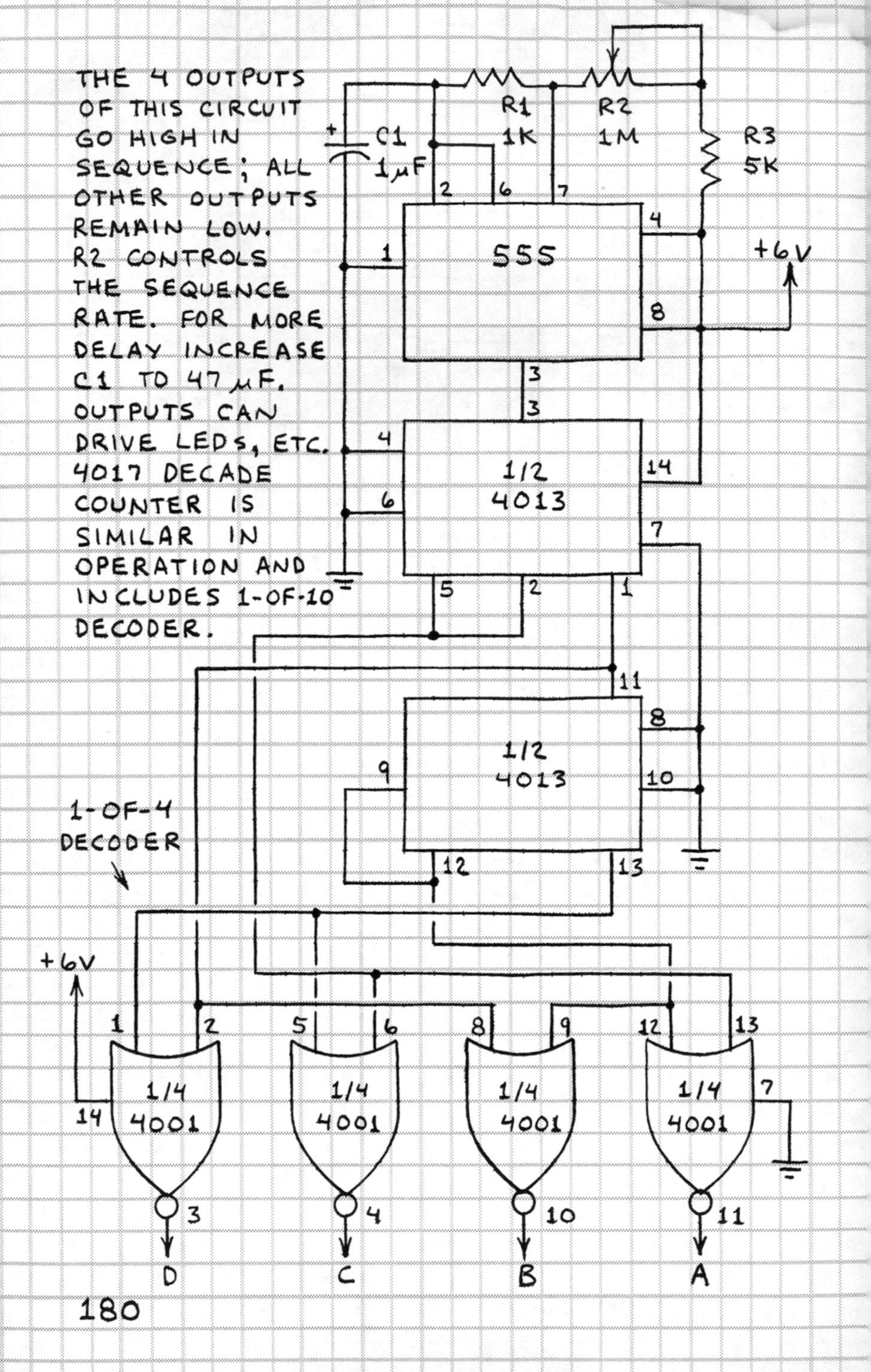

# SHIFT REGISTER

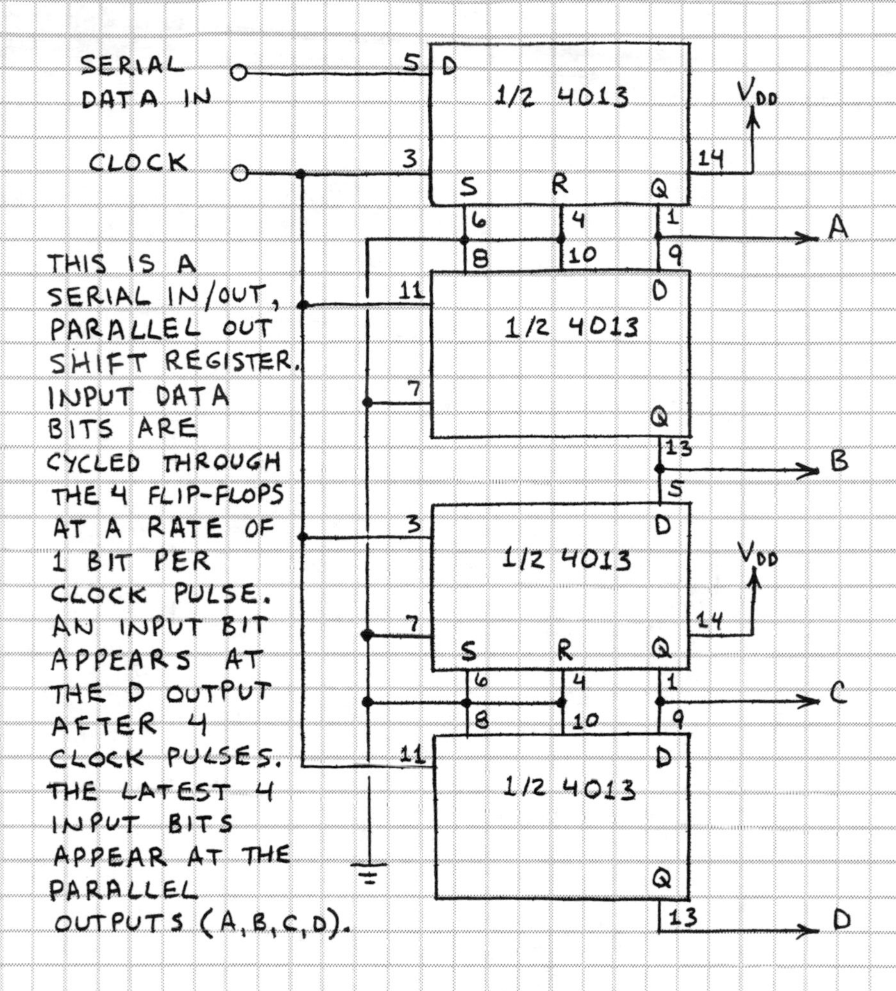


# COUNT TO N AND RECYCLE

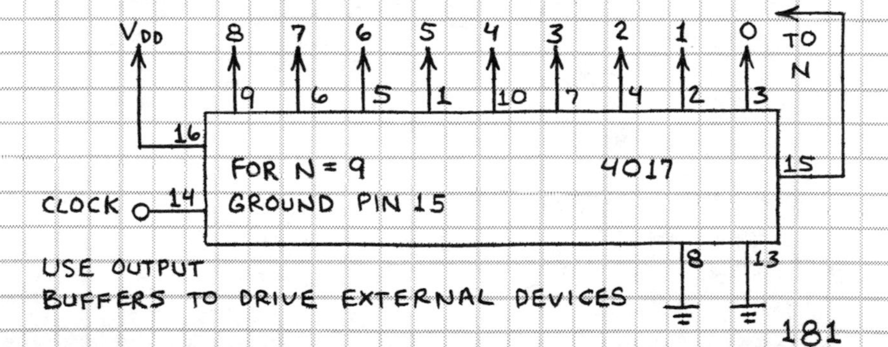

# PROGRAMMABLE TIMER

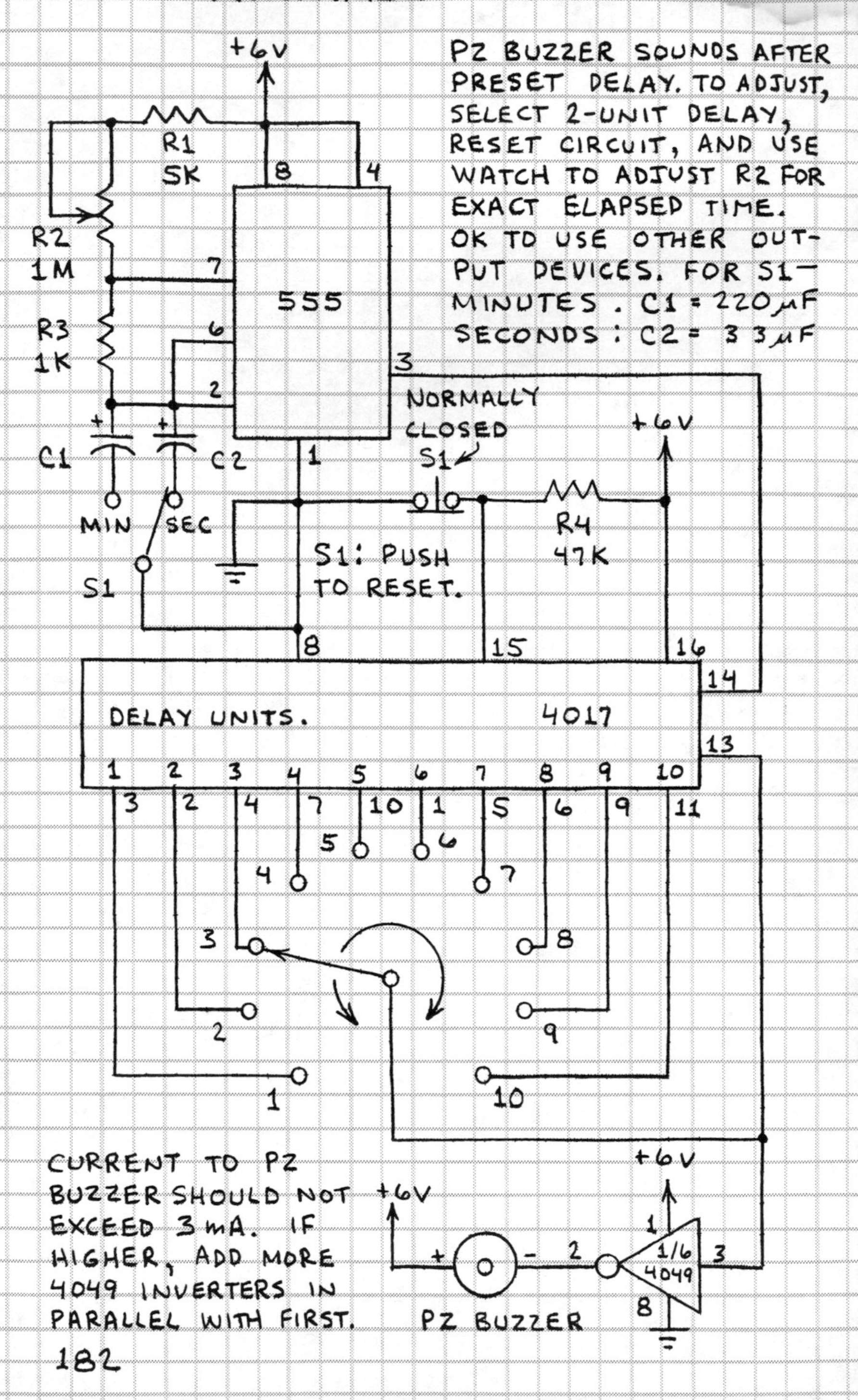


PZ BUZZER SOUNDS AFTER PRESET DELAY. TO ADJUST, SELECT 2-UNIT DELAY, RESET CIRCUIT, AND USE WATCH TO ADJUST R2 FOR EXACT ELAPSED TIME. OK TO USE OTHER OUTPUT DEVICES. FOR S1— MINUTES . C1 = 220 μF SECONDS : C2 = 3 3 μF

CURRENT TO PZ BUZZER SHOULD NOT EXCEED 3 mA. IF HIGHER, ADD MORE 4049 INVERTERS IN PARALLEL WITH FIRST.

# RANDOM NUMBER GENERATOR

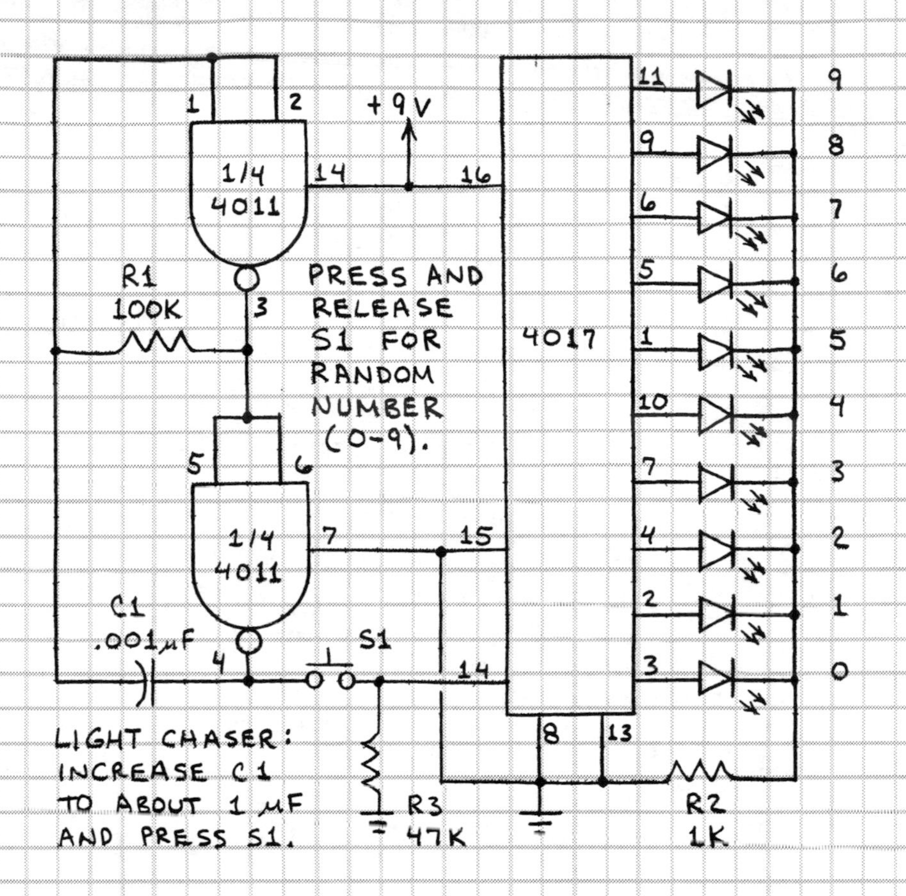


# DIVIDE-BY-TWO COUNTERS

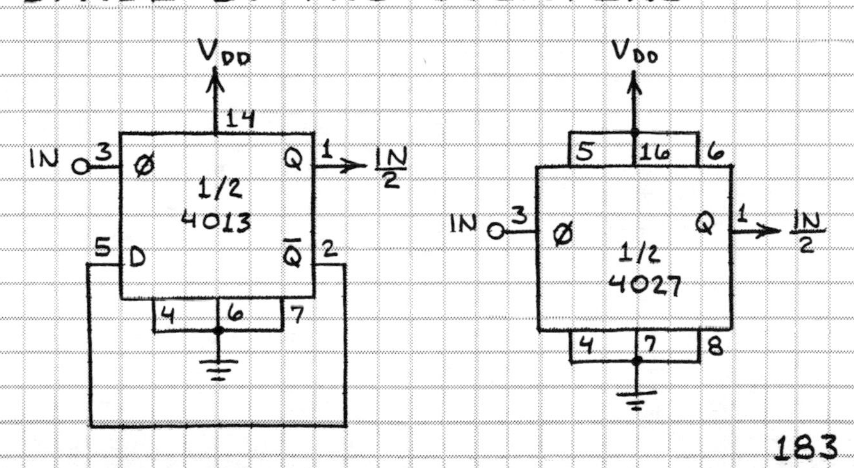

# LOGIC FAMILY INTERFACING

THESE GUIDELINES PERMIT TTL AND CMOS LOGIC CIRCUITS TO BE INTERCONNECTED.

## TTL TO TTL

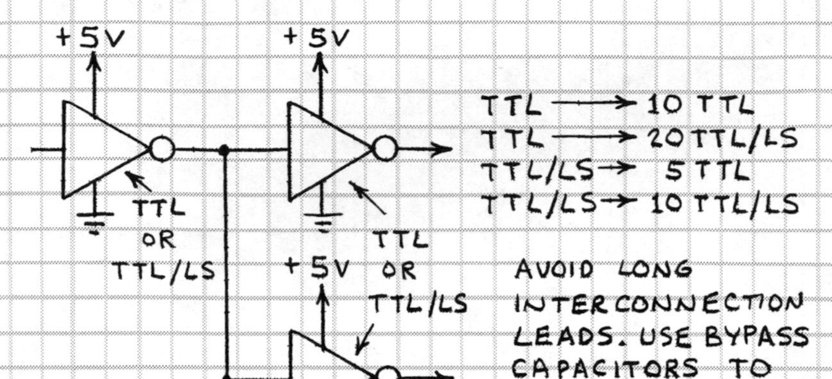


AVOID LONG INTERCONNECTION LEADS. USE BYPASS CAPACITORS TO DECOUPLE POWER SUPPLY NOISE (P.10).

## TTL TO CMOS

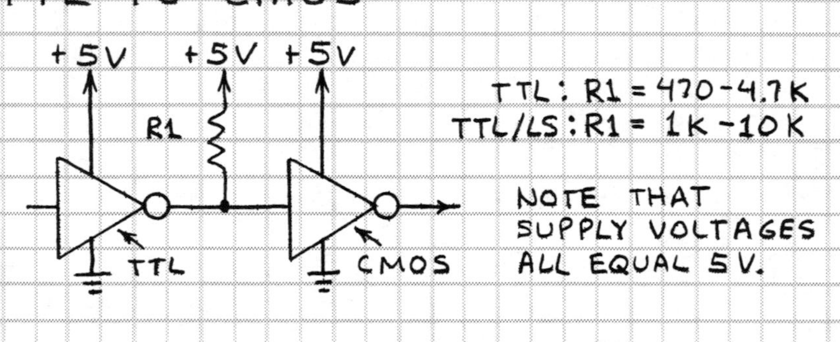


NOTE THAT SUPPLY VOLTAGES ALL EQUAL 5V.

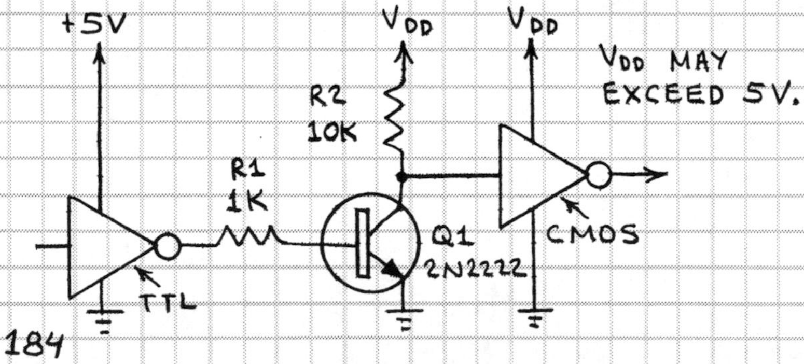


$V_{DD}$ MAY EXCEED 5V.

# CMOS TO CMOS

A CMOS GATE OUTPUT CAN DRIVE UP TO 50 CMOS INPUTS. AVOID LONG INTERCONNECTIONS AND CONNECT ALL UNUSED INPUTS TO $V_{DD}$ OR GROUND.

# CMOS TO TTL

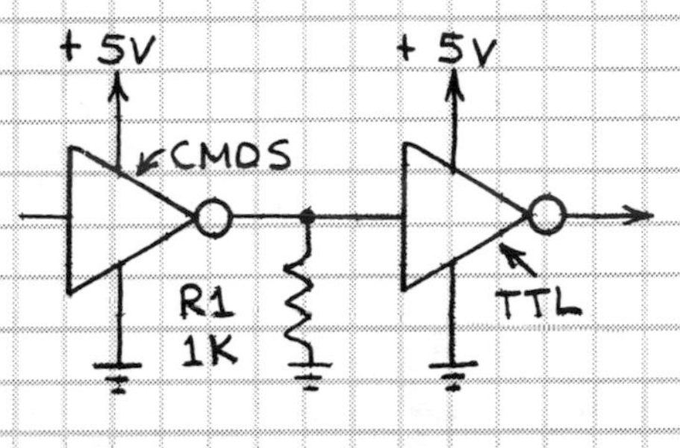


IF POSSIBLE, CHECK LOGIC INTERFACES WITH A LOGIC PROBE TO MAKE SURE THEY WORK AS INTENDED.

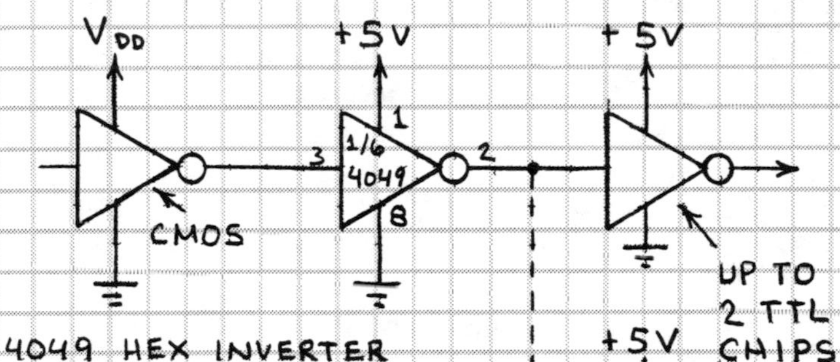


4049 HEX INVERTER IS DESIGNED FOR INTERFACING.

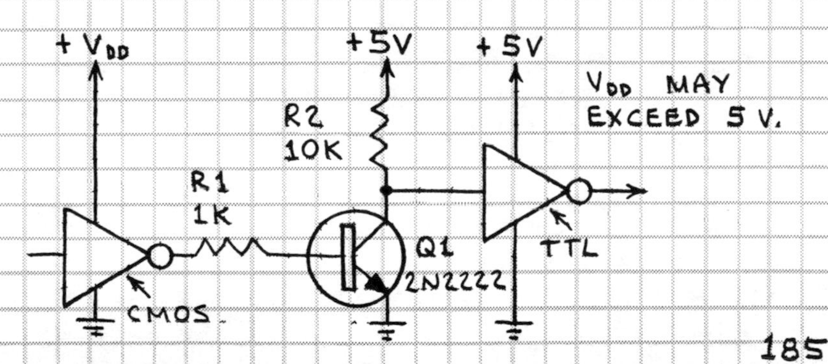


$V_{DD}$ MAY EXCEED 5 V.

# DIGITAL LOGIC TROUBLESHOOTING

SOMETIMES A DIGITAL LOGIC CIRCUIT MAY FAIL TO OPERATE OR MAY OPERATE IMPROPERLY. THE TROUBLESHOOTING PROCEDURES GIVEN HERE WILL ENABLE THE SOURCE OF MOST PROBLEMS TO BE IDENTIFIED. A LOGIC PROBE IS VERY HELPFUL WHEN TESTING A LOGIC CIRCUIT. USE A COMMERCIAL UNIT OR BUILD YOUR OWN.

1. REMOVE POWER FROM THE CIRCUIT.

2. CHECK ALL WIRING CONNECTIONS.

3. ARE ANY CHIP PINS BENT AND NOT FULLY INSERTED IN THE SOCKET OR CIRCUIT BOARD?

4. ARE ALL SOLDER CONNECTIONS GOOD?

5. DO ALL INPUTS GO SOMEWHERE? EVEN INPUTS OF UNUSED CMOS GATES MUST GO TO $V_{DD}$ OR GROUND.

6. DOES THE CIRCUIT OBEY ALL OPERATING REQUIREMENTS (SUPPLY VOLTAGE, ETC.)?

7. DOES THE CIRCUIT INCLUDE DECOUPLING CAPACITORS CLOSE TO AND ACROSS THE SUPPLY PINS OF EVERY FEW CHIPS?

8. ARE THE INPUTS AND OUTPUTS OF ALL LOGIC CHIPS PROPERLY INTERFACED?

IF THESE STEPS DO NOT ISOLATE THE SOURCE OF THE PROBLEM, ONE OR MORE LOGIC CHIPS MAY BE DEFECTIVE. REMEMBER THAT CMOS CHIPS ARE ESPECIALLY VULNERABLE TO STATIC ELECTRICITY AND IMPROPER INPUT AND OUTPUT LOADING. FINALLY, BE SURE THE POWER SUPPLY WORKS PROPERLY AND IS CAPABLE OF PROVIDING SUFFICIENT CURRENT TO THE CIRCUIT IT POWERS.

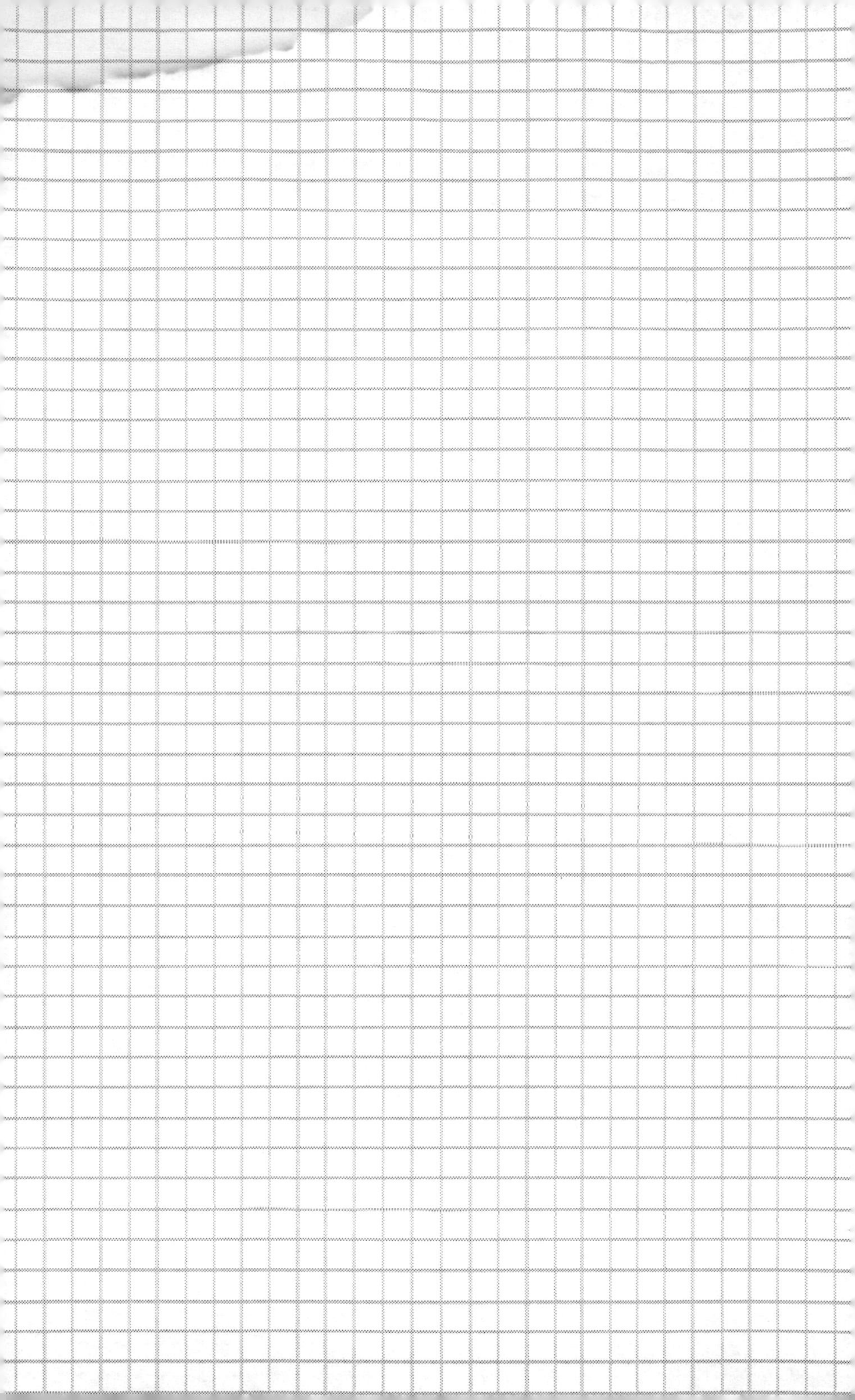